Wissenschaftliche Beiträge
zur Medizinelektronik

Band 4

Wissenschaftliche Beiträge zur Medizinelektronik

Band 4

Herausgegeben von

Prof. Dr. Wolfgang Krautschneider

Stimulation of Neurons by Electrical Means

Vom Promotionsausschuss der
Technischen Universität Hamburg-Harburg
zur Erlangung des akademischen Grades
Doktor-Ingenieur (Dr.-Ing.)

genehmigte Dissertation

von
Mario Alberto Meza Cuevas

aus
Guadalajara, Mexiko

2015

1.Gutachter: Prof. Dr.-Ing. Wolfgang H. Krautschneider

2.Gutachter: Prof. Dr.-Ing. Hoc Khiem Trieu

Tag der mündlichen Prüfung: 28.08.2015

Mario A. Meza-Cuevas

Stimulation of Neurons by Electrical Means

Logos Verlag Berlin

λογος

Wissenschaftliche Beiträge zur Medizinelektronik

Herausgegeben von
Prof. Dr. Wolfgang Krautschneider

Technische Universität Hamburg-Harburg
Institut für Nano- und Medizinelektronik
Eißendorfer Str. 38
D-21073 Hamburg

Bibliografische Information Der Deutschen Bibliothek

Die Deutsche Bibliothek verzeichnet diese Publikation in der Deutschen Nationalbibliografie; detaillierte bibliografische Daten sind im Internet über http://dnb.ddb.de abrufbar.

ISBN 978-3-8325-4152-1
ISSN 2190-3905

Logos Verlag Berlin GmbH
Comeniushof, Gubener Str. 47,
10243 Berlin

Tel.: +49 (0)30 / 42 85 10 90
Fax: +49 (0)30 / 42 85 10 92
http://www.logos-verlag.de

Acknowledgments

I would like to thank the Prof. Krautschneider and the Dr. Schroeder, who during my stay at the institute gave me guidance in the technical, scientific and strategic areas related to each point treated in this work.

My generation colleagues, with whom I had deep technical discussions and a lot of long journeys working together, and became great friends: Lait Abu-Saleh, Bryce Bradford and Farzard Hosseini. The old guys of the institute, because they made an effort to transfer the acquire knowledge. To Ricardo Starbird, for his support in the chemical areas treated in this document, and for his friendship.

To Fabian Wagner, who as a colleague of the institute shared his knowledge, and later together with Martin Kadner became my commercial partner.

To the staff of the institute, who support us (the PhDs) in all the activities: Ronald Mielke, Ute Schmidt, Gabrielle Heinrich, Silke Bade and Volker Stradtmann.

To The Institute of Optical and Electronic Materials, and the Institute of Microsystems Technology, who provided equipment and facilities for various activities such as developing electrodes and wire bonding of silicon substrates.

Nevertheless, the project of the dissertation was not only work of 6 years, but an academic, professional and personal formation. During my life, I have received support and encouragement of many people.

I thank my mother who gave me the life, she died in 2003, I dedicate this work to her. My father gave me a good education and taught me the fundamentals of this wonderful profession. My sisters, they are always supporting me unconditionally.

To the family Sánchez García, who has been a second family to me and gave me their trust, love and friendship.

To all my friends in Mexico, including their families, they are always supporting me in everything. Ricardo Rivera a good friend whom I will always have in my heart.

Joel Gallo and Laura Schneider, they helped me to move to Hamburg and to arrange all the related issues. It is not easy to start a new life in another city. Thus, I would like to thank to all the friends I have made in Germany, because they gave me their support, especially to my three first friends in Hamburg: Fabina Dietrich, Marie-Luise Schneider and Jessica Boensch. At the end but not least, my flatmate, Mareen Schwarze, she became in the last 4 years another sister to me.

To my teachers at the University of Guadalajara and the CINVESTAV, because they have contributed to my professional formation.

To the CONACYT and DAAD who awarded me a scholarship for the first three years of my PhD.

Summary

Medicine has reached a great number of achievements. It is well known that in most cases these achievements have not only been obtained through the application of medical science, but also through engineering in the design and development of instruments to assess diseases, and tools to improve surgery techniques. Engineering has now assumed another useful role in developing electronic devices that emulate some functions of the human body.

Electronic implants can stimulate the proper nerve cells to restore affected functions. The task of performing stimulation of neurons by applying electrical pulses is called electrical neurostimulation. This is useful to restore vision in humans, to restore hearing through cochlear implants and to assist physiological functions.

The implants must fulfill certain characteristics; they should be small enough for the implantation; they are usually powered by an inductive coupling and the energy is limited. Low-power considerations thus play an important role in the design of biomedical systems.

Smaller electrodes are required in order to stimulate small groups of neurons or even single neurons. But smaller electrodes present higher impedances that result in higher voltages, i.e. higher power consumption.

To support power efficiency and the use of advanced transistor technologies, the operation voltages should be as small as possible. Similarly, it is crucial to maintain low power dissipation, since the increase of temperature in cells could be harmful.

In this dissertation, the problematic will be partly approached by using different disciplines. We demonstrate our achievements by using an alternative material for the electrodes, PEDOT, which has emerged as the most promising of the electrically conducting polymers because of its electrical properties, its improvement in charge injection and the possibility of chemical modification to enhance biocompatibility.

Stimulation pulse has to be optimized for maximum power efficiency. Through simulations with equivalent electrical models, we analyze the use of different waveforms for electrical neurostimulation. We then show the development of a portable stimulator, which is useful to enable experimentation with several stimulation parameters. We then corroborate the results using *In Vivo* experimentation.

We also show the evolving design of a fully implantable stimulator. The stimulator is a low-power design that is able to drive microelectrodes small enough to perform invasive electrical stimulation. It is also capable of delivering several waveforms to save energy during the stimulation. It presents a schema for charge-balanced stimulation in invasive stimulation, because an imbalanced stimulation may cause neural tissue damage. The ASIC is a scalable multi-channel neurostimulator for applications requiring larger number of stimulation sites, such as implants to restore vision.

Today, this application is of great interest due to vision loss among a significant number of persons worldwide on account of degenerative diseases. Patients can regain some visual perception by artificial electrical stimulation on one point of the visual pathway.

Contents

Introduction

Around 1780 the Italian physician Luigi Galvani (9 September 1737 – 4 December 1798) accidentally touched the exposed nerve of a frog with a metal scalpel, which picked up a charge. There were sparks and the dead frog's leg kicked as though alive. He continued experimenting with human cadavers and came to reject the old theory of Descartes, which posited that nerves were pipes carrying fluids. Galvani established that the nerves were electrical conductors. This was not entirely true but he helped to establish the basis for electrophysiology, which is the science of electrical properties of biological cells and tissues.

The mechanism of natural stimulation of neurons is performed by electrochemical process by releasing neurotransmitters. However, it is also possible to stimulate them by artificial ways, which means applying an external electrical signal to excite the neurons by inducing a flow of ions through the neuronal cell membrane. The fact of stimulating a neuron through an electrical impulse causes the firing of an action potential between its membranes. The action potential propagates along its axon to its synaptic ending to release neurotransmitters and so stimulate a neighbor neuron or even muscle by natural process.

The nervous system is basically composed of three kinds of neurons: sensory neurons, interneurons and motor neurons.

The sensory neurons carry information from receptors to the brain. The receptors are located all over the body to give human sense (sight, hearing, taste, smell and touch). The sense in the human is the ability of perceiving the environment through the conversion of several physical or chemical stimuli to electrical signals.

The motor neurons carry information from the brain to the target. The targets are mostly muscles that are stimulated by signals coming from the brain. Some muscles are responsible for voluntary movements, such as locomotion, or involuntary movements, such as physiological functions (respiration, circulation, etc).

The task of performing stimulation of neurons by applying electrical pulses is called electrical neurostimulation [1].

Electrical neurostimulation is useful to improve sensory deficits such as restoring vision [2] through retinal [3][4][5][6], optical nerve [7] [8], lateral geniculate nucleous (LGN)[9], or cortical implants [10]. It is also useful for restoring hearing through cochlear implants [11].

Electrical neurostimulation has been effective in reducing symptoms of some neurological disorders by applying vagus nerve stimulation (VNS) in cases of epilepsy [12] and depression [13]; by employing deep brain stimulation (DBS) in the case of Parkinson´s [14], epilepsy [15], depression [16], or dystonia [17]; or also by spinal cord stimulation for alleviating some types of chronic pain [18].

Neurological disorders, like spinal cord injury, multiple sclerosis, cerebral palsy or stroke, can partially or completely disable people in their freedom of movement. Damage to the central nervous system or particular nerve pathways can lead to severe consequences and a failure or disorder of muscles can occur. The use of electronic devices allows the neural pathways to be bypassed from central nervous system, and to directly stimulate the motor neurons which are electrically excitable and produce muscle contraction. The neuromuscular electrical stimulation (NMES), also called functional neuromuscular stimulation (FNS), has been used to restore the movement of extremities [19] [20] [21] [22], and it is also used for rehabilitation [23] [24], support in daily life, for alleviating some types of chronic pain or muscle training [25].

The electrical neurostimulation is also useful for assisting physiological functions through electronic implants such as the pacemaker [26], bladder prosthesis [27] and the phrenic pacer [28] [29].

Medicine has achieved a great deal in the treatment of many diseases and injuries. It is well known that in most cases these achievements have not only been obtained through the application of medical science, but also through engineering in the design and development of instruments to assess diseases, and tools to improve surgery techniques.

Nowadays, engineering plays another useful in developing electronic devices that emulate some functions of the human body. Electronic implants or prostheses can perform several tasks such as acting like a sensor to measure the content of any chemical in the blood, delivering any pharmaco, monitoring cardiac rhythm and brain activity or even stimulating the proper nerve cells to restore physiologically affected functions.

Electronic engineering has up to now made very important advances. For instance, fifty years ago a whole room was needed to store a computer, whereas now computers are small enough to fit in a volume of a few cubic centimeters. All of this has been thanks to the miniaturization and great integration of a large amount of components in an integrated circuit using micro and nanoelectronics so that more powerful and smaller electronic devices, such as electronic implants, can be built.

The circuitry of a neurostimulator is an electronic design that is capable of generating either current or voltage signals and is mainly composed of blocks represented in the figure below:

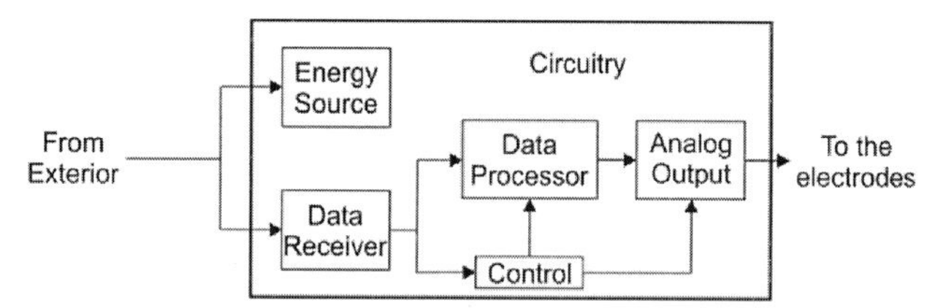

Figure 1 Block diagram of the modules involving a neurostimulator.

The system should fulfill certain characteristics regarding physical design: small systems are preferable due to the necessity to carry the external device or the implantation of the system into the human body. Due to the size and location of the devices, they are powered by batteries or by wireless methods such as radio frequency. The energy transfer is thus limited. Low-power considerations therefore play an important role in the design of biomedical systems.

The data receiver should receive either commands or information from the exterior. For implants with experimental purposes, this task could be performed by wired methods. However, this could cause infection at the skin level. Thus, for fully implantable devices wireless methods are preferred, like in the energy source module. In fact, the power and information could be transmitted in the same way.

The control is a required module capable of receiving instructions from the outside. It interprets them to command the rest of the elements in the design at the proper time in order to deliver the stimulating signals.

The data processor, depending on the design, could be composed of different elements such as serial to parallel converter, registers with the purpose of digital storage for analog signals, or even a signal generator depending on the desired waveform.

At the stage of analog output, digital information is converted into analog signals to perform the stimulation. These signals could be either current or voltage signals, though the most commonly used are current signals since the natural stimulation is performed through electrical current. Besides, current pulses are preferred over voltage pulses in order to eliminate variations in the stimulation threshold as a result of changes in the electrode-tissue impedance.

This stage should be able to deliver the required charge in order to stimulate the target neuron. For this purpose, the desired current output as well as the pulse width should be considered as these parameters are related to the strength-duration curve for stimulation of a neuron. Not less important is the required voltage to reach such current as it depends on the impedance of the electrodes.

The neurostimulators are classified as follows: external neurostimulators use surface electrodes and the device is external; partial implantable neurostimulators are when electrodes are implanted and the device is external; fully implantable neurostimulators are when both the electrodes and the device are implanted. In any case, they should establish a bioelectrical interface with the body. The interface, called electrodes, has to meet certain electrical and biocompatible requirements. The material of the electrodes must be carefully chosen for a proper bioelectrical interface without disturbing the cells and tissues [30] [31] [32]. In the case of invasive stimulation, smaller electrodes are required in order to stimulate small groups of neurons or even single neurons. But smaller electrodes present higher impedances that result in higher voltages, i.e. higher power consumption.

To support power efficiency and the use of advanced transistor technologies, the operation voltages should be as small as possible. Similarly, it is important to maintain low power dissipation, since the increase in temperature in tissues or the brain could be harmful.

Chapter one of this dissertation will present the scientific background of the nervous system, which is essential to understand the electrical stimulation mechanism. The problematic will then be partly approached by using different disciplines.

In chapter two, the interface created between the electrode surface and the biological tissue is described. We then show our achievements by using an alternative material for the electrodes, Poly-Ethylenedioxythiophene (PEDOT), which has emerged as the most promising of the electrically conducting polymers (ECP) due to its electrical properties, its improvement in charge injection and the possibility of chemical surface modification in enhancing biocompatibility.

Stimulation pulse has to be optimized for maximum power efficiency. Thus, in chapter three, by performing simulations with a passive membrane model, we show that energy can be saved by performing stimulation with non-rectangular waveforms. Afterwards, in order to achieve more accurate results, an NMES equivalent electrical circuit was developed, which we used to perform simulations with different waveforms. The results show that non-rectangular waveforms can provide more energy-efficient neural stimulation. Therefore, we developed a portable stimulator that enables experimentation with several stimulation parameters. Using our stimulator, we then performed *In Vivo* tests to corroborate the simulations. The experiments were conducted by applying transcutaneous electrical muscle stimulation to different subjects.

In chapter four we describe our evolving design of a fully implantable stimulator. The design consists of three application specific integrated circuits (ASICs), each one an enhanced version of the previous one. The stimulator is a low power design, able to drive microelectrodes small enough to perform invasive electrical stimulation. It is also capable of delivering several waveforms to save energy during the stimulation. It presents a schema for charge-balanced stimulation in invasive stimulation because an imbalanced stimulation may cause neural tissue damage. The third version is a scalable multi-channel neurostimulator for applications requiring a larger number of stimulation sites, such as implants to restore vision. This application is nowadays of great interest due to vision loss among a significant number of persons worldwide on account of degenerative diseases. Patients can regain some visual perception by artificial electrical stimulation on one point of the visual pathway.

Chapter five presents the conclusions of each chapter of this work.

1. Physiological Background

A theoretical background is shown in this chapter, the function and physiology of the nervous system, as well as the neuron, which is necessary to understand the neurostimulation elicited by electrical means. The information of this chapter is basically based on [33] [34].

1.1. The Nervous System

The nervous system is subdivided anatomically into the Central Nervous System (CNS) and the Peripheral Nervous System (PNS). CNS comprises the brain and spinal cord. PNS is all of the nervous system that is not part of the CNS, including the nerves emerging from the brain (called cranial nerves) and the spinal cord (called spinal nerves), see Figure 1.1.1. The spinal nerves exit the spinal canal at the level of their respective vertebrae and transmit the information between the brain and peripheral nervous system.

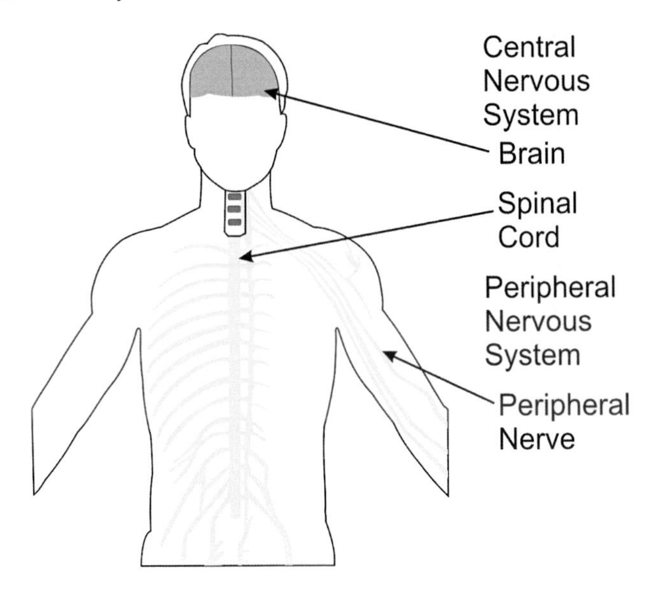

Figure 1.1.1 The nervous system.

The peripheral nerves convey neural messages from: the sense organs and sensory receptors inward to the CNS; and from the CNS outward to the muscles and glands of the body. The PNS is composed by nerves, the afferent nerves carry nerve impulses from receptors toward the CNS, and efferent nerves carry nerve impulses away from the CNS to effectors.

The structure of a nerve with its parts is shown in Figure 1.1.2. The axons are surrounded by a layer of connective tissue called the endoneurium. The axons are bundled together into groups called fascicles, each fascicle is wrapped in a layer of connective tissue called the perineurium. The entire nerve is wrapped in a layer called the epineurium.

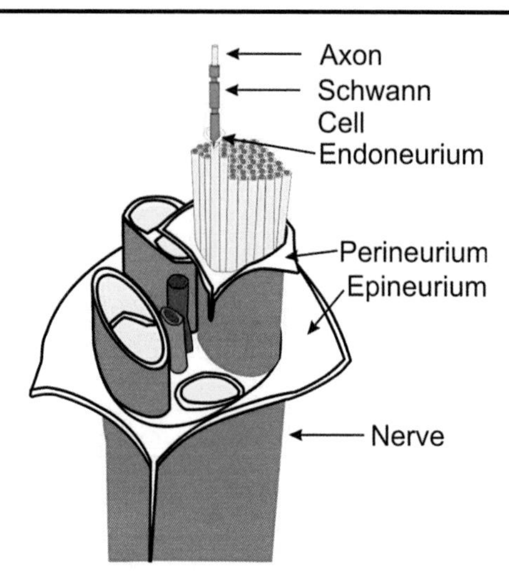

Figure 1.1.2 The structure of the nerve.

1.1.1. The Neuron

The term neuron was coined by the German anatomist, Heinrich Wilhelm Gottfried von Waldeyer-Hartz (6 October 1836 – 23 January 1921). The neuron is the basic structure unit of the nervous system on mammals. The neuron is an electro-chemically excitable cell responsible for processing and carrying information from and into the brain. There are different types of neurons, however most of them have the same parts as the typical motor neuron illustrated in Figure 1.1.3. The cell body (soma) contains the nucleus and the rest of the organelles (mitochondria, endoplasmatic reticulum, ribosomes, among others) and is the metabolic center of the neuron. Dendrites are branches of a neuron, they extend from the cell body and ramify to create a network. The dendrites receive impulses from other cells and transfer them to the cell body (afferent signals). The effect of these impulses may be excitatory or inhibitory. A cortical neuron may receive impulses from tens or even hundreds of thousands of neurons. The neurons have typically a long fiber that originates from a somewhat thickened area of the cell body, it is called the axon hillock. The first portion of the axon is called the initial segment. The axon transfers the signal from the cell body to another nerve or to a muscle. The axon furcates into presynaptic terminals, each ending in synaptinc knobs which are also called terminal buttons or boutons. They contain granules or vesicles in which the synaptic transmitters secreted by the nerves are stored.

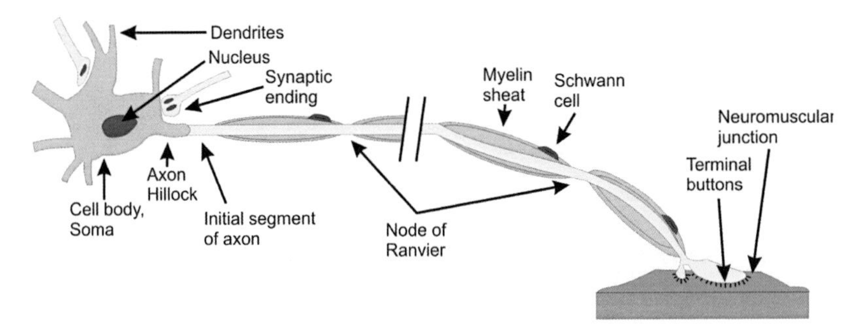

Figure 1.1.3 Motor neuron with a myelinated axon.

The axons of many neurons are myelinated, i.e. they have a sheath of myelin around the axon, the myelin is a protein-lipid complex. In the peripheral nervous system, myelin forms when a Schwann cell wraps its membrane around an axon up to 100 times. The myelin sheath is covering the axon except at its ending and at the nodes of Ranvier. The nodes of Ranvier are periodic 1 μm gaps with an interdistance of about 1 mm. The insulating function of myelin is explained in the section 1.1.8. Not all neurons are myelinated; they are simply surrounded by Schwann cells without the wrapping of the Schwann cell membrane that produces myelin around the axon.

Based on the function, the neurons can be classified as afferent or sensory neurons, they convey information from tissues and organs into the CNS; efferent or motor neurons, they transmit signals from the central nervous system to the effector cells; interneurons, they connect neurons within specific regions of the CNS. These types of neurons are shown in the Figure 1.1.4, normally the sensory neurons are bipolar or pseudo-unipolar cells, they have two specialized processes: a dendrite that carries information to the cell and an axon that transmits information from the cell. The motor neuron and interneuron are multipolar cells, they have one axon and many dendrites. The typical dimensions and number of elements on the nervous system are shown in Table 1.1.1,[35].

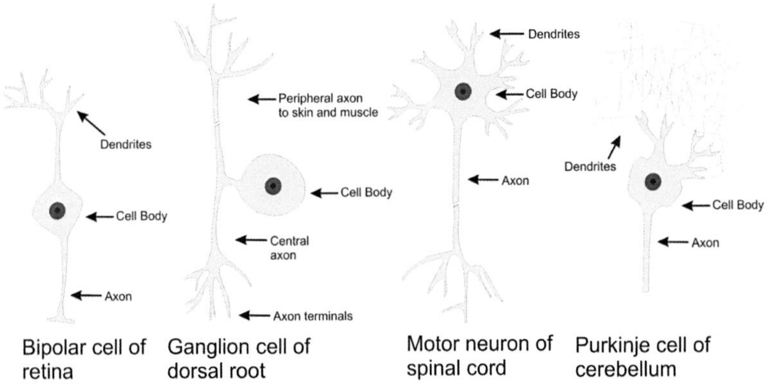

Figure 1.1.4 Type of neurons.

Table 1.1.1 Typical dimensions and number of elements on the nervous system [35].

	Value	Unit
Node of Ranvier length	4	µm
Axon diameter	1-20	µm
Soma diameter	5-20	µm
Nerve fascicle diameter	1	mm
Nerve diameter	Several	mm
Number of auditory fibers	3×10^4	
Number of optical fibers	10^6	
Number of brain neurons	10^{11}	
Connections per brain cell	10^4	

1.1.2. Axonal Transport

Neurons are secretory cells. The secretion duct is normally close to the cell body, the neurons differ in that, they have it generally at the end of the axon. The organelles responsible for synthesizing the proteins are located in the cell body, and then transported along the axon. The transport of proteins and polypeptides to axonal ending is carried by axoplasmic flow. Axonal transport is essential to the neuron growth and survival. Thus, the cell body maintains the functional and anatomic integrity of the axon. The transport occurs along microtubules (made of tubulin) that run along the length of the axon. Two elements are required for the transportation, dynein and kinesin, which are molecular motors that move cargoes in both directions of the axon, Figure 1.1.5. Orthograde (also called anterograde) transport moves molecules from the cell body to the axon terminals. Orthograde transport can be fast and slow; fast axonal transport occurs at about 400 mm/day, and slow axonal transport occurs at 0.5 to 10 mm/day. Retrograde transport is in the opposite direction, from the synapse to the cell body. Some synaptic vesicles are recycled in the membrane, and some used synaptic vesicles are carried back to the cell body by fast retrograde transport, it occurs at about 200 mm/day. Retrograde transport is mediated by dynein, and informs the soma of conditions at the axon terminals.

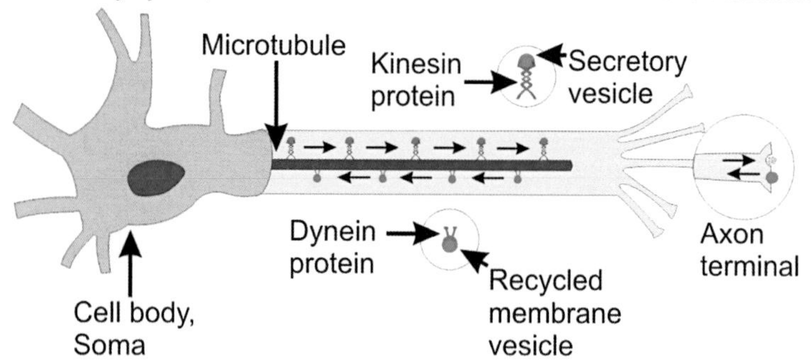

Figure 1.1.5 Axonal transport.

1.1.3. The Synapse

Information from the cell body to the adjacent cell travel along the axon, and then across the synapse. The synapse word comes from the Greek “synapsis” and it means conjunction. The synapse is a structure that passes an electrical or chemical signal to another neuron.

In a chemical synapse, Figure 1.1.6, electrical activity in the presynaptic terminal is converted by the activation of voltage-gated calcium channels. Then, the neurotransmitter (chemical) is released to the postsynaptic terminal. The presynaptic terminal is the part of the synapse that is on the side of the axon; the part on the side of the adjacent cell is called the postsynaptic terminal. Between these terminals exists a gap, the synaptic cleft, it has a thickness of 10 - 50 nm. The synapse between a motor nerve and the muscle is called the neuromuscular junction and it innervates.

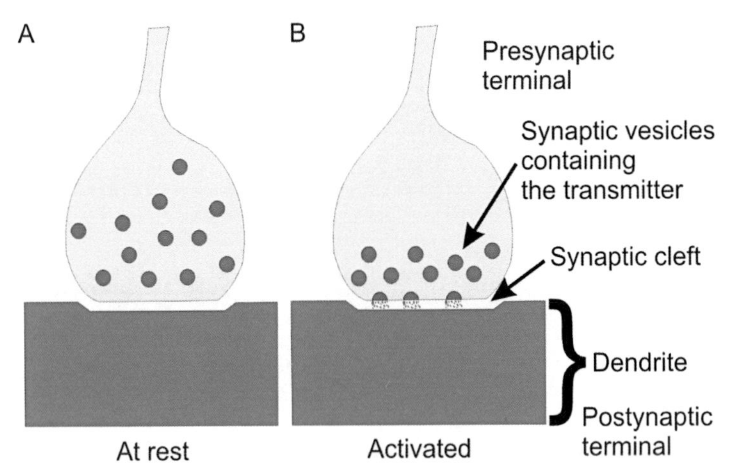

Figure 1.1.6 Chemical synapse, a) the synaptic vesicles contain a chemical transmitter, b) the activation reaches the presynaptic terminal and the transmitter is released activate the postsynaptic membrane.

In an electrical synapse, the presynaptic and postsynaptic cell membranes are connected by channels called gap junctions that permit current to flow passively. The current flow changes the postsynaptic membrane potential. The main advantage of an electrical synapse is the rapid transfer of signals from one cell to the next.

1.1.4. Excitation of the Neuron

Nerve cells have a threshold for excitation. The stimulus may be electrical, chemical or mechanical. The transmembrane voltage changes when a nerve cell is stimulated. After the stimulation the membrane voltage returns to its original resting value. The stimulation can be excitatory or inhibitory. Excitatory stimulation, depolarizing, is characterized by a change of the potential inside the cell relative to the outside in the positive direction, it means, the negative resting voltage decreases; Inhibitory stimulation, hyperpolarizing, is characterized by a change in the potential inside the cell relative to the outside in the negative direction, it means, the magnitude of membrane voltage increases.

If the stimulus is insufficient to cause the transmembrane potential to reach the threshold, then the membrane will not activate. However, if the excitatory stimulus is strong enough to reach the transmembrane threshold, the membrane produces an electric impulse, which is called the action potential. The action potential fails to occur if the stimulus is subthreshold in magnitude, and it occurs with constant amplitude and form regardless of the strength of the stimulus if the stimulus is at or above threshold intensity, the action potential is therefore “all or none” in character and is said to obey the all-or-none law.

The activation of a cell depends on the strength and duration of the stimulus. A proper stimulation can cause the membrane activation with a short and strong stimulus or a long and weak stimulus. This characteristic is called the strength-duration curve, Figure 1.1.7. The smallest current that initiates the activation is called the rheobasic current or rheobase. The rheobasic current needs theoretically an infinite duration to trigger the membrane activation. The time needed to excite the cell with twice rheobase current is called chronaxy.

Accommodation is the adaptation of the cell to repetitive stimulus, and its excitation threshold rises. Facilitation increases the excitability of the cell, i.e. there is a decrease in excitation threshold.

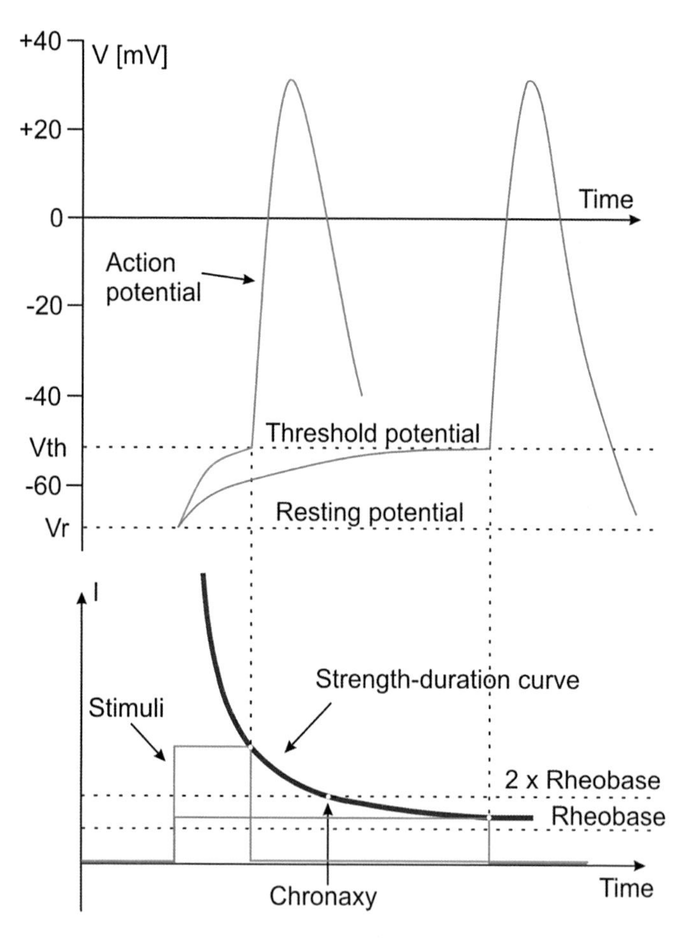

Figure 1.1.7 The strength-duration curve.

1.1.5. Resting Membrane Potential

If two electrodes are placed on the axon, one on the surface and the other is inserted into the interior of the cell, a constant potential difference is observed, the inside more negative than the outside. A membrane potential results from separation of positive and negative charges across the cell membrane. In neurons, the resting membrane potential is usually about – 70 mV.

The resting membrane potential is because of an unequal distribution of ions of one or more species between the interior and exterior of the membrane, i.e. concentration gradient. In order to keep this distribution difference the membrane must be permeable to these ion species. The permeability is provided by channels in the bilayer membrane, these channels are usually permeable to a single species of ions.

1.1.6. Ionic Fluxes During the Action Potential

The difference of the membrane potential is explained because of the different concentration of ions between its inside and outside. In neurons, the concentration of potassium (K^+) is much higher inside than outside the cell, while the reverse is the case for sodium (Na^+). Transmembrane ion channels allow ions into or out of cells. The cell membrane of nerves contains different types of ion channels. The channels can be voltage-gated or ligand-gated. The electrical events in nerves are explained by the behavior of these channels.

The channels conductance changes during the action potential are depicted in Figure 1.1.8. It is necessary to remember that the conductance of an ion is the reciprocal of its electrical resistance in the membrane, the ion conductance is related to the membrane permeability to that ion. Some voltage-gated Na^+ channels become active when a depolarizing stimulus is present. And when the threshold potential is reached, all the ion channels open, the Na^+ channels exceed the K^+ and other channels and an action potential occurs. The membrane potential changes, it moves toward the equilibrium potential for Na^+ (+ 60 mV) but does not reach it during the action potential, primarily because the Na^+ channel opening is short-lived. The Na^+ channels enter to an inactivated state and stay there just for a few milliseconds, they return then to the resting state. Since the membrane potential is reversed during the overshoot, the direction of the electrical gradient for Na^+ is reversed and this limits Na^+ influx.

The repolarization occurs because of the opening of voltage-gated K^+ channels. The opening of these channels is slower and more prolonged than the opening of the Na^+ channels. Thus, the increase in K^+ conductance comes after the increase in Na^+ conductance. The movement of positive charge out of the cell due to the K^+ channels helps to complete the process of repolarization. The hyperpolarization is explained because of the slow return of the K^+ channels to the original state. Thus, the K^+ channels bring the action potential to an end and cause their gates close.

The membrane is insensitive to new stimuli when activation has been initiated, this phase is called the absolute refractory period. The cell can be activated again from the relative refractory period, it is near to the end of the action potential, during this phase a stimulus stronger than normal is necessary.

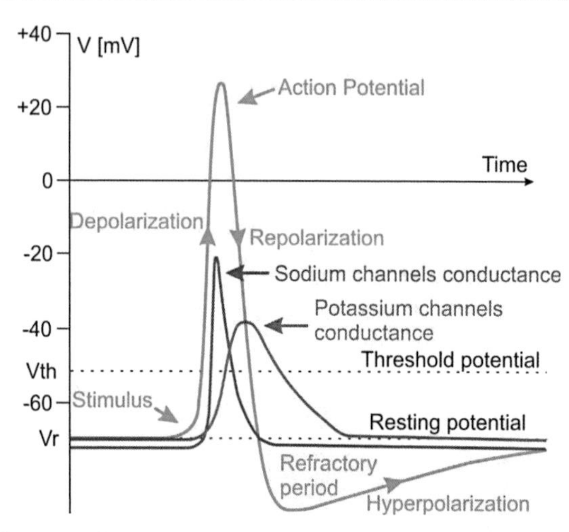

Figure 1.1.8 The changes in the membrane during the action potential.

1.1.7. Distribution of Ion Channels in Myelinated Neurons

The ion channels in the axon play an important role in the initiation and regulation of the action potential. Voltage-gated Na^+ channels are highly concentrated in the nodes of Ranvier and the initial segment in myelinated neurons. The activation pulses are normally generated in the initial segment and in the first node of Ranvier. And during the saltatory conduction (next section), the pulses jump to the other nodes of Ranvier.

The number of Na^+ channels per square micrometer of membrane in myelinated mammalian neurons has been estimated to be 50-75 in the cell body; 350-500 in the initial segment; less than 25 on the surface of the myelin; 2,000-12,000 at the nodes of Ranvier; and 20-75 at the axon terminals. Along the axons of unmyelinated neurons, the number is about 110. In many myelinated neurons, the Na^+ channels are enclosed by K^+ channels that are involved in repolarization.

1.1.8. Conduction of the Axon

The action potential propagates in an axon as an unattenuated nerve impulse. The potential difference between excited and unexcited regions of an axon cause small local currents. They flow from the excited areas toward the unexcited areas. Thus, the flow stimulates the unexcited regions.

Despite stimuli can be seen in the dendrites or some, activation originates normally in the soma. Activation in form of the action potential is first seen in the root of the axon. It propagates from there along the axon. If artificial stimuli is applied somewhere along the axon, the action potential propagates in both directions from the stimulus site.

The conduction in myelinated axons is up to 50 times faster than in the unmyelinated axons. This is because of the salutatory conduction in a myelinated axon, See Figure 1.1.9. Conduction in myelinated axons occurs in a circular current flow pattern. Myelin sheets are an effective insulator, and current flow through it is negligible.

Thus, depolarization in myelinated axons jumps from one node of Ranvier to the next, with the current sink at the active node serving to electrotonically depolarize the node ahead of the action potential to the firing level.

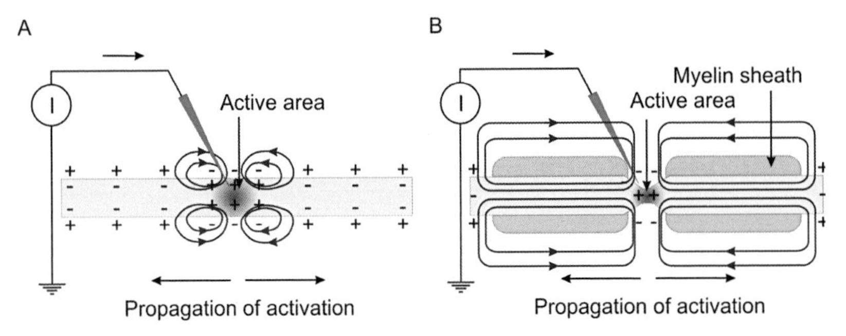

Figure 1.1.9 Conduction of the axon, a) continuous conduction in an unmyelinated axon, b) saltatory conduction in a myelinated axon.

The electric properties and the geometry of the axon determine its conduction velocity. The change of sodium conductance due to activation is an important property of the membrane. If the maximum value achieved by the sodium conductance increases, the sodium ion current and rate of change in the membrane voltage increase as well. The result is a higher gradient of voltage, increased local currents, faster excitation, and increased conduction velocity. The decrease in the threshold potential facilitates the triggering of the activation process.

The capacitance of the membrane determines the amount of charge required to achieve an action potential and therefore affects the time needed to reach the threshold. Large capacitance values mean a slower conduction velocity.

Another important parameter for the conduction velocity is the resistivity of the medium, since it also affect the depolarization time constant. The resistance is directly proportional to the time constant and inversely proportional to the conduction velocity.

The temperature also affects the time constant of the sodium conductance; a decrease in temperature decreases the conduction velocity.

The above effects are reflected in an expression derived by Muler and Markin (1978)

Equation 1.1
$$v = \sqrt{\frac{\mathrm{i}_{Na\,max}}{r_i c_m^2 V_{th}}}$$

Where v = velocity of the action potential [m/s], $\mathrm{i}_{Na\,max}$ = maximum sodium current per unit length [A/m],V_{th} =threshold voltage [V], r_i = axial resistance per unit length [Ω/m] , c_m = membrane capacitance per unit length [F/m].

The membrane capacitance per unit length of a myelinated axon is smaller than in an unmyelinated axon. Thus, the myelin sheath increases the conduction velocity.

The resistance of the axoplasm per unit length is inversely proportional to the cross-sectional area of the axon, it is important to remember that the relationship between the diameter and the sectional area is exponential. The membrane capacitance per unit length is directly proportional to the diameter. The time constant is the product of the resistance and capacitance. Thus, an increment in the axon diameter decreases the resistance more than the increase of the capacitance. The conduction velocity of myelinated axons is therefore directly proportional to its diameter.

1.1.9. Hodgkin-Huxley Model

The Hodgkin-Huxley model is a mathematical model, which uses electric elements to describe the behavior of the Na^+ and K^+ channels in the membrane. The behavior of each channel is voltage- and time-dependent. The model is useful to describe how action potentials in neurons are initiated and propagated.

The electrical equivalent circuit is illustrated in Figure 1.1.10. The capacitor represents the capacitance of the cell membrane; the two variable resistors represent the voltage-dependent Na^+ and K^+ conductances; the fixed resistor represents a voltage-independent leakage conductance and the three voltage sources represent reversal potentials for the corresponding conductances. The arrow labelled "I" represents an external electrical stimulation [36].

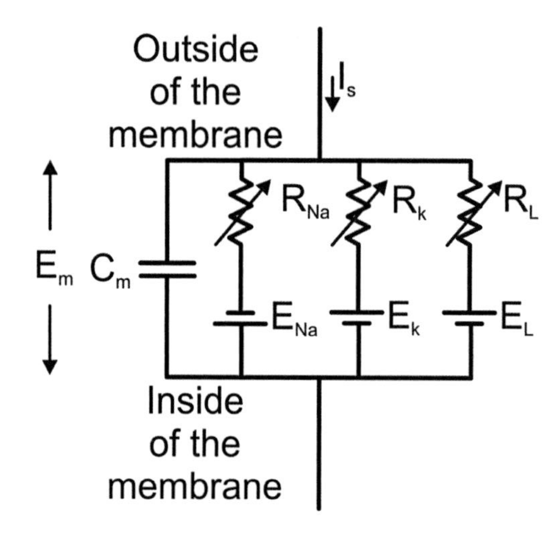

Figure 1.1.10 Hodgkin-Huxley electrical circuit representing the membrane.

2. Electrodes

The electrodes are the interface between the circuitry and the biological tissue. The biological tissue is rich in electrolytes. Some reactions at the electrode-tissue interface are required to mediate the transition from electron flow in the electrode to ion flow in the tissue. The interface created in the electrode surface is described in section 2.1.

In the section 2.2, we show our achievements by using an alternative material for the electrodes, PEDOT, which has emerged as the most promising of the electrically conducting polymers (ECP), because of its electrical properties, its improvement in charge injection and the possibility of chemical surface modification in order to enhance the biocompatibility.

2.1. Electrode-Electrolyte Interface

When a metal electrode is placed inside an electrolyte such as extracellular fluid, an interface is immediately formed between the two phases. Both of the metal and solution are electroneutral until electrons are transferred across the interface. As a consequence, an electric field is formed that influences the chemical. It inhibits the reduction reaction, while accelerating the oxidation reaction. These reactions eventually reach an equilibrium condition. Then, the currents due to electrons transfer to and from the metal are equal. At equilibrium, the current density flowing across the interface in both directions results in a net current of zero.

The electric field generated by these electron transfer reactions also has an impact on the electrolyte. Water dipoles orient themselves in the field in a layer at the metal surface forming a hydration sheath. After the hydration sheath is generally a layer of ions, see Figure 2.1.1. The plane in which the ions get collected is called the Outer Helmholtz Plane (OHP) [37].

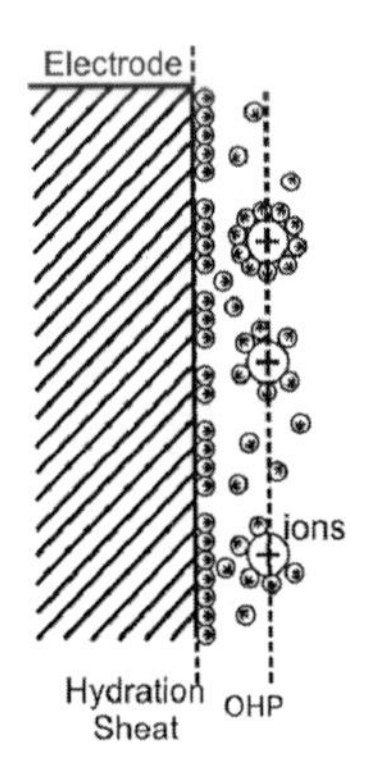

Figure 2.1.1 The hydration sheath and the outer Helmholtz plane.

Between the OHP and the hydration sheath, a larger number of solvent dipoles (with random orientations) may be present. In some cases, the solute ions might replace some of the water molecules in the hydration sheath. In such cases, the ions would be in direct contact with the metal. This case arises only if the total energy associated with the system would be lowered due to the breaking of the sheath around the ion, the transport of the ion from the OHP to the solid surface and due to the adsorption reaction at the solid. The plane in which the ions that are in contact with the electrode is shown in Figure 2.1.2, it is called the Inner Helmholtz Plane (IHP) [37].

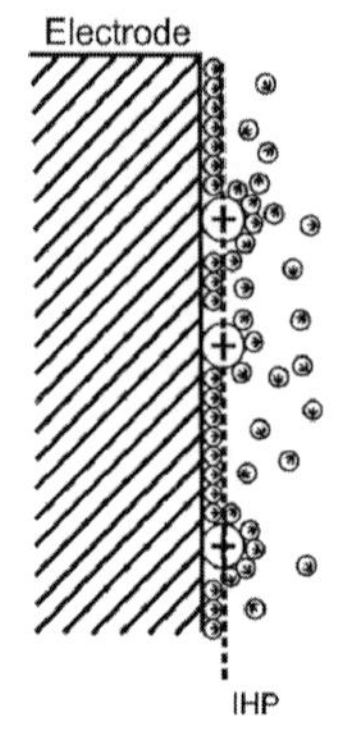

Figure 2.1.2 The inner Helmholtz plane.

2.1.1. Interface Model Elements

The space between the electrode and the outer neural membrane is a highly complex region, different models have been proposed to model this electrochemical region with equivalent circuits, however the basic elements used are: double layer capacitance, Warburg element, charge transfer resistance and spreading resistance.

Double Layer Capacitance

It is expected that, a charged surface in contact with an electrolyte solution attracts ions of opposite charge and repels ions of same charge, thus establishing an ion environment in the immediate vicinity of the surface. Two parallel layers of charge are formed, the charge on the surface itself and the layer of oppositely charged ions near the surface [38].

The first and simplest model of the charge separation at interfaces was proposed by Helmholtz. The initial theory assumed that, the charge of solvated ions was confined to a rigid sheet. The interface was predicted to behave like a simple capacitor, with the oriented water dipole as dielectric. The capacitance is given in Equation 2.1, and is determined by the permittivity of electrolyte $\epsilon_o \epsilon_r$, the area of the interface A_{el} and the distance between the electrode and electrolyte d.

Equation 2.1
$$C_H = \epsilon_o \epsilon_r \frac{A_{el}}{\mathrm{d}}$$

In the Helmholtz model, the electric potential falls from its surface value to zero in the bulk solution over the thickness of the layer of counter ions, Figure 2.1.3.

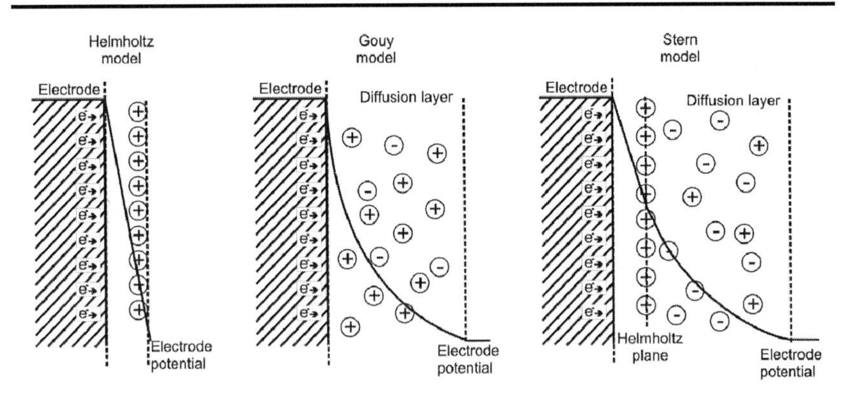

Figure 2.1.3 Models of the electric double layer.

The simple model of Helmholtz neglected the capacitance dependence on potential. Gouy and Chapman modified the simple Helmholtz model, they predicted a dependence of the measured capacitance on both potential and electrolyte concentration. According to Gouy and Chapman, ions in the electric double layer are subjected to electrical and thermal fields. This allows the Maxwell-Boltzmann statistics to be applied to the charge distribution of ions as a function of distance away from the metal surface akin to the distribution of negatively charged ions surrounding a positive ion [39].

Gouy and Chapman suggested that the ions which neutralize the surface charge are spread out into solution, forming the diffuse double layer, where he potential falls slowly to the bulk solution value [38], Figure 2.1.3.

The differential capacitance of the diffuse double layer is given by Equation 2.2 [39]. It is valid at low electrolyte concentrations in the vicinity of the point of zero charge [40].

Equation 2.2 $$C_G = \sqrt{\frac{n_0 \epsilon e_0^2}{2\pi kT}} \cdot \cosh\left(\frac{e_0 V_0}{2kT}\right)$$

Where n_0 is the bulk concentration of ion in solution and V_0 is the potential applied to the electrode.

The Gouy-Chapman model represented substantial improvement over the Helmholtz model in that a dependence of the differential capacitance on both potential and concentration was predicted. In this diffuse layer, the net charge density decreases with distance from the phase boundary. However it only gives acceptable results in very dilute solution at potentials close to the potential of zero charge, but incorrect in a physiological environment, as the measured capacitance is much lower than that calculated from equation.

Stern combined the Helmholtz model and Gouy-Chapman model and proposed a new model to describe the interface capacitance. Stern assumed the charge on the solution side resided partially in a compact layer (Helmholtz capacitance), and the remainder charges diffuse along the distance (Gouy-Chapman diffusion capacitance), this two capacitors are connected in series, Figure 2.1.3. The capacitance is given by:

Equation 2.3 $$\frac{1}{C_S} = \frac{1}{C_H} + \frac{1}{C_G}$$

Grahame proposed that, although the immediate layer in contact with the electrode is occupied by solvent molecules, it may be possible for some ionic or uncharged species to penetrate into this region. This could occur if the ion had no solvation shell, or if the solvation shell was lost when the ion came close to the electrode. Ions in direct contact were expected to be specifically adsorbed [41]. Three regions in the electrode-electrolyte interface are proposed in this model, as illustrated in Figure 2.1.4.

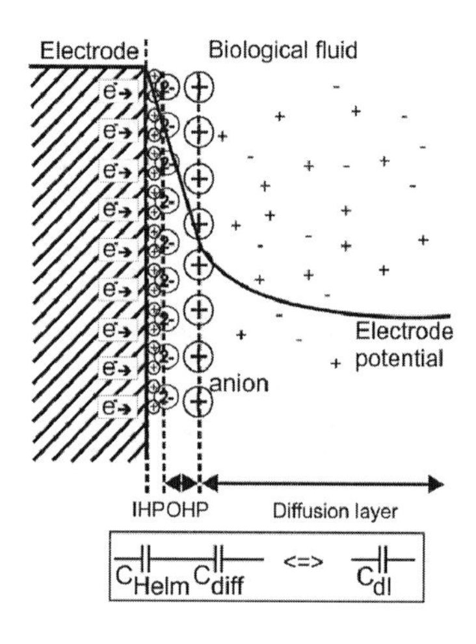

Figure 2.1.4 The double layer.

The IHP extends from the electrode to a plane passing through the centers of the specifically adsorbed ions. The OHP passes through the centers of hydrated ions at their distance of closes approach to the electrode. Diffusion layer, the ions are not tightly confined with increasing distance from the electrode. Potential changes linearly with distance up to the OHP and then exponentially through the diffuse double layer region. For this model the capacitance is given by:

Equation 2.4 $$\frac{1}{C} = \left(\frac{1}{C_{IHP}} + \frac{1}{C_{OHP}}\right) + \frac{1}{C_G}$$

Where, C_{IHP} and C_{OHP} are the integral capacitances of the space between the electrode and the IHP, and between the IHP and the OHP.

Equation 2.5 $$C_{H1} = \epsilon_o \epsilon_{IHL} \frac{A_{el}}{d_{IHL}}$$

Equation 2.6 $$C_{H2} = \epsilon_o \epsilon_{OHL} \frac{A_{el}}{d_{OHL} - d_{IHL}}$$

Warburg Element

When the current density of the electrolyte is increased, free diffusion of reactants is inhibited, in other words, frequency dependent impedance exists in the interface, and it counteracts with the diffusion effect. This impedance is first proposed by Warburg.

The impedance is inversely proportional to frequency. At high frequencies the Warburg impedance is small since diffusing reactants do not have to move very far. At low frequencies the reactants have to diffuse further, thereby increasing the Warburg impedance. The Warburg diffusion element, Equation 2.7, is a Constant Phase Element (CPE) with a constant phase of 45° (phase independent of frequency) and with a magnitude inversely proportional to the square root of the frequency:

Equation 2.7 $$Z_W = \frac{\sigma_W}{\sqrt{\omega}} - j\frac{\sigma_W}{\sqrt{\omega}}$$

The parameter σ_W in the equation is called Warburg coefficient, it is defined by:

Equation 2.8 $$\sigma_W = \frac{RT}{n^2F^2A_{el}\sqrt{2}}\left(\frac{1}{C^o\sqrt{D_o}} + \frac{1}{C^r\sqrt{D_r}}\right)$$

Where R is the gas constant, T is the absolute temperature, n is the number of electrons transferred, F is the Faraday constant, C^o and C^r are the surface concentration of oxidant and reductant species in the bulk, and D_o and D_r are the diffusion coefficients of the oxidant and reductant.

However, it is valid if the diffusion layer has an infinite thickness. If the diffusion layer is bounded, then it can be used the Equation 2.9, called finite Warburg.

Equation 2.9 $$Z_W = \frac{\sigma_W}{\sqrt{\omega}} - j\frac{\sigma_W}{\sqrt{\omega}}\tanh\left(\delta\sqrt{j\frac{\omega}{D}}\right)$$

Where, δ is the Nernst diffusion layer thickness and D the average value of the diffusion coefficients of the diffusing species.

Charge Transfer Resistance

The current that crosses the electrode-electrolyte interface experiences a charge transfer resistance. At equilibrium state of the electrode-electrolyte interface, there is a constant flow of charge across the interface, but the net flow is zero. The absolute value of this current is known as the exchange current density, i_o. The charge transfer resistance R_{ct} represents the Faradaic process where charges transfer between the electrode and electrolyte by means of oxidation-reduction reactions.

Equation 2.10 $$R_{ct} = \frac{RT}{nFi_o}$$

The expression is valid when the electrode overpotential is small. Otherwise, the Butler-Volmer equation can be used to calculate the current depending on applied voltage.

Spreading Resistance

The resistance measured between the working electrode and the reference electrode is another component for the interface equivalent circuit. If we assume that the reference electrode is infinitely large and the working electrode is in direct contact with the electrolyte, the spreading resistance of square electrode with side length l is given by the Equation 2.11, and the resistance of circular electrode with radius r is given by the Equation 2.12, [42].

Equation 2.11 $$R_s = \frac{\rho \ln 4}{\pi l}$$

Equation 2.12 $$R_s = \frac{\rho}{4r}$$

2.1.2. Interface Models

With those elements between the electrode-electrolyte interface, many equivalent circuit models have been proposed. The most popular models used for Electrochemical Impedance Spectroscopy (EIS) are the simple model, Randles model and Geddes Beker Model.

Simple Model

In the simple model, the interface is modelled by a series connected double layer capacitor and spreading resistor, Faradaic process and diffusion effects are omitted. It is suitable for modelling the electrode-electrolyte interface at high frequency and immersed in electrolyte with low concentration. The equivalent circuit is shown in Figure 2.1.5

Figure 2.1.5 The simple model.

In Figure 2.1.6, the behavior of the model is shown. A basis RC circuit does not allow low AC frequency and DC current to pass through, as the capacitance C presents a very high impedance to the current and rejects the current at low frequency where $\omega \to 0$. At high frequency the resistor R_s limits the flow of the current. The time constant for the circuit is $\tau = R_s C_{dl}$. The plot reveals the resistive component R_s and corresponding phase angle 0° at high frequency. At $f < f_c = 1/(2\pi R_s C_{dl})$ the impedance decreases proportionally to the frequency increase. This circuit is representative of a conductive solution with conductance inversely proportional to the R-parameter: At low frequencies a "blocking" fully capacitive interface emerges where only the double layer charging represented by the capacitor C_{dl} is presented. Total impedance of the circuit and its real and imaginary components can be expressed as [43]:

Equation 2.13 $$Z = R_s - j\frac{1}{\omega C_{dl}}$$

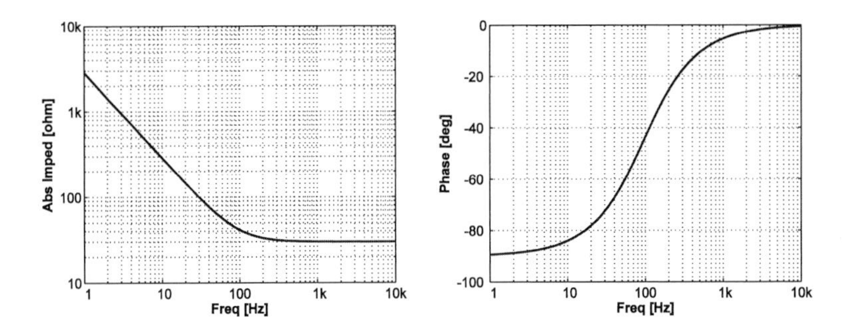

Figure 2.1.6 The behavior of the simple model.

Randles Model
If the Faradaic process is taken into consideration, the charge transfer resistance will be included in the model, it is parallel connected with the double layer capacitor. This model evolves from the research of J. Randles. The equivalent circuit is illustrated in Figure 2.1.7.

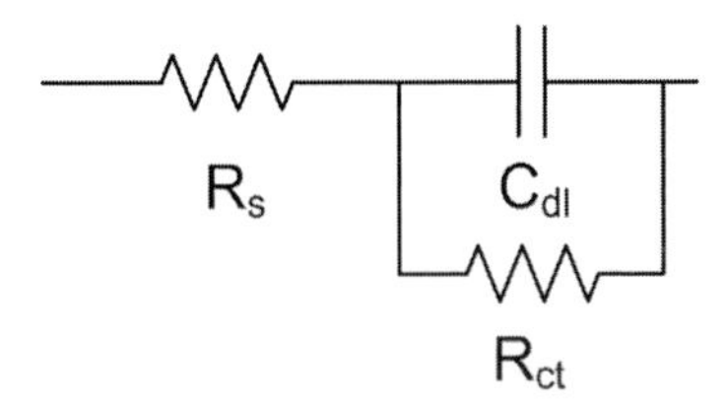

Figure 2.1.7 The Randles model.

A Randles circuit is one of the simplest and most common cell models used for many aqueous, conductive, and ionic solutions. It includes only solution resistance R_s, a parallel combination of double layer capacitor C_{dl}, and a charge transfer resistance R_{ct}.

The behavior of the model is shown in Figure 2.1.8. The Randles circuit is characterized by a high-frequency purely resistive solution component where the phase angle is 0° and the current is $I = V/R_s$. At lower frequencies where the phase angle approaches 0° and the current approaches $I = V/(R_s + R_{ct})$. At medium frequencies a corresponding increase in the absolute value of the phase angle and the finite charge transfer resistor R_{ct} is present. The time constant for the circuits $\tau = R_{ct}C_{dl}$, with the imaginary impedance reaching maximum absolute value at the critical frequency $f_c = 1/(2\pi R_{ct}C_{dl})$. The simplified Randles cell model is often the starting point for other more complex model, its impedance is given by [43]:

Equation 2.14 $$Z = R_s + \frac{R_{ct}}{1+(\omega R_{ct}C_{dl})^2} - j\frac{\omega R_{ct}^2 C_{dl}}{1+(\omega R_{ct}C_{dl})^2}$$

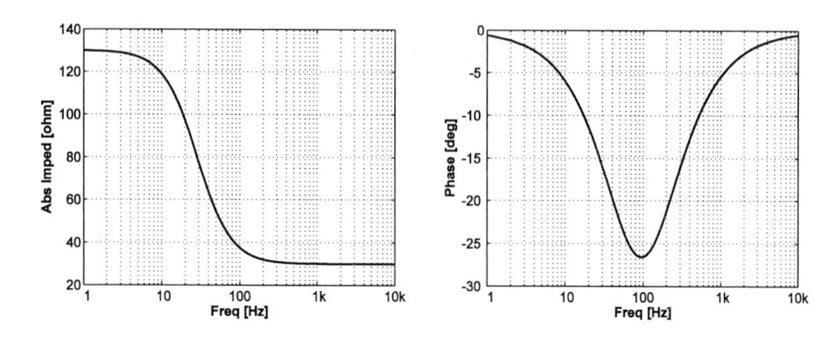

Figure 2.1.8 The behavior of the Randles model.

Full Randles Model

When the frequency related parameter Warburg element is considered, the diffusion effect will be better described with the circuit shown in Figure 2.1.9. The Warburg impedance is an element, with constant phase angle of 45°. The circuit is sometimes called full Randles model. This circuit models a cell where polarization is due to a combination of kinetic and diffusion processes [44].

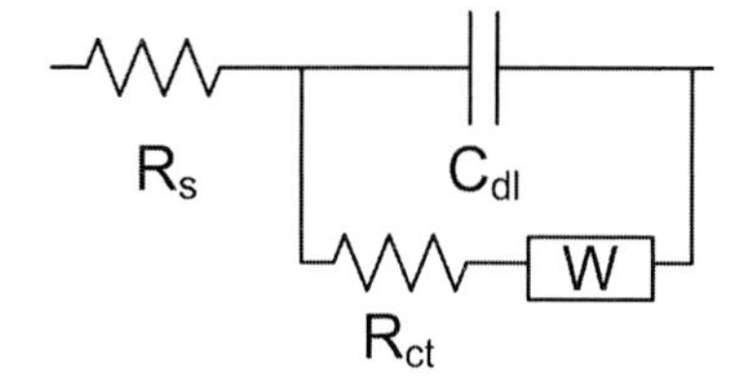

Figure 2.1.9 The full Randles model.

2.1.3. Charge Transfer

In the metal electrode phase and in attached electrical circuits, charge is carried by electrons. In the physiological medium, or in more general electrochemical terms the electrolyte, charge is carried by ions, including sodium, potassium, and chloride. The central process that occurs at the electrode-electrolyte interface is a transduction of charge carriers from electrons in the metal electrode to ions in the electrolyte. The information of this section is basically based on [45] [46].

There are two primary mechanisms of charge transfer at the electrode-electrolyte interface, illustrated in Figure 2.1.10.

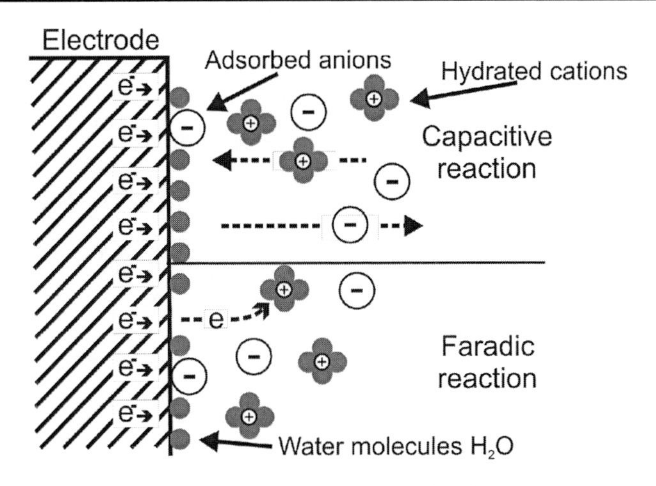

Figure 2.1.10 Electrode-Electrolyte Interface.

"One is a capacitive (non-Faradaic) reaction, where no electrons are transferred between the electrode and electrolyte"[46]. They involve the charging and discharging of the electrode-electrolyte double layer. "Non-Faradaic reactions include redistribution of charged chemical species in the electrolyte" [46].

"Capacitive charging can be either electrostatic, involving purely double-layer ion-electron charge separation and electrolyte dipole orientation, or electrolytic, involving charge stored across a thin, high-dielectric-constant oxide at the electrode-electrolyte interface" [45].

"The second mechanism is a Faradaic reaction in which electrons are transferred between the electrode and electrolyte, resulting in reduction or oxidation of chemical species in the electrolyte" [46].

Faradaic reactions are further divided into reversible and nonreversible Faradaic reactions. Reversible Faradaic reactions include those where the products either remain bound to the electrode surface or do not diffuse far away from the electrode. In an irreversible Faradaic reaction, the products diffuse away from the electrode.

"For noble metal electrodes, primarily Pt and PtIr alloys, the Faradaic reactions are confined to a surface monolayer, and these reactions are often described as pseudocapacitive, although electron transfer across the interface still occurs. Charge can also be stored and injected into tissue from valence changes in multivalent electrode coatings that undergo reversible reduction-oxidation (redox) reactions. These coatings, which are typified by iridium oxide, but include newer materials such as PEDOT, are mixed conductors, exhibiting both electron and ion transport within the bulk of the coating. The three-dimensional structure of the coatings provides more charge for stimulation, but access to this charge is limited by the rate of electron and ion transport within the coating" [45]. The classification of the most used material for electrodes is shown in Table 2.1.1.

Table 2.1.1 Most used material for electrodes and its transfer mechanism [45].

Material	Mechanism
Pt and PtIr alloys	Faradaic/Capacitive
Activated iridium oxide	Faradaic
Thermal iridium oxide	Faradaic
Sputtered iridium oxide	Faradaic
Tantalum/Tantalum oxide	Capacitive
Titanium nitride	Capacitive
PEDOT	Faradaic

“PEDOT is an ECP that, like iridium oxide, exhibits both electronic and ionic conductivity” [45]. “Whether PEDOT injects charge by capacitive or Faradaic processes depends on the potential range over which the electrode is driven during pulsing. For potential more positive than -0.6 V Ag-AgCl, capacitive charge injection may dominate, although the detail of the charge injection processes in PEDOT at high current densities have not been fully explored” [45].

Capacitive Interface

“In principle, capacitive charge-injection is more desirable than Faradaic charge-injection because no chemical species are created or consumed during a stimulation pulse. Because the double-layer charge per unit area at an electrode-electrolyte interface is small, high charge-injection capacity is only possible with capacitor electrodes that are porous or employ high dielectric constant coatings. Both strategies have limitations” [45].

“If only non-Faradaic redistribution of charge occurs, the electrode-electrolyte interface may be modeled as a simple electrical capacitor called the double-layer capacitor C_{dl}. This capacitor is formed due to several physical phenomena. First, when a metal electrode is placed in an electrolyte, charge redistribution occurs as metal ions in the electrolyte combine with the electrode. This involves a transient transfer of electrons between the two phases, resulting in a plane of charge at the surface of the metal electrode, opposed by a plane of opposite charge, as counterions, in the electrolyte. A second reason for formation of the double layer is that some chemical species such as halide anions may specifically adsorb to the solid electrode, acting to separate charge. A third reason is that polar molecules such as water may have a preferential orientation at the interface, and the net orientation of polar molecules separates charge” [46].

“If the net charge on the metal electrode is forced to vary (as occurs with charge injection during stimulation), a redistribution of charge occurs in the solution. Consider two metal electrodes immersed in an electrolytic salt solution. A voltage source is applied across the two electrodes so that one electrode is driven to a relatively negative potential and the other to a relatively positive potential. At the interface that is driven negative, the metal electrode has an excess of negative charge. This will attract positive charge (cations) in solution toward the electrode and repel negative charge (anions). In the interfacial region, there will be net electroneutrality, because the negative charge excess on the electrode surface will equal the positive charge in solution near the interface. The bulk solution will also have net electroneutrality. At the second electrode, the opposite processes occur, i.e., the repulsion of anions by the negative electrode is countered by attraction of anions at the positive electrode. If the total amount of charge delivered is sufficiently small, only charge redistribution occurs, there is no transfer of electrons across the interface, and the interface is well modeled as a simple capacitor. If the polarity of the applied voltage source is then reversed, the direction of current is reversed, the charge redistribution is reversed, and charge that was injected from the electrode into the electrolyte and stored by the capacitor may be recovered" [46].

Faradaic Interface

"Charge may also be injected from the electrode to the electrolyte by Faradaic processes of reduction and oxidation, whereby electrons are transferred between the two phases. Reduction, which requires the addition of an electron, occurs at the anode, while oxidation, requiring the removal of an electron, occurs at the cathode" [46]. Faradaic charge injection results in the creation of chemical species, which may either go into the solution or remain bound to the electrode surface. Unlike the capacitive charge injection mechanism, if these Faradaic reaction products diffuse sufficiently far away from the electrode, they cannot be recovered upon reversing the direction of current.

Reversible and Irreversible Faradaic Reactions

"There are two limiting cases that may define the net rate of a Faradaic reaction. At one extreme, the reaction rate is under kinetic control; at the other extreme, the reaction rate is under mass transport control. For a given metal electrode and electrolyte, there is an electrical potential (voltage) called the equilibrium potential where no net current passes between the two phases. At electrical potentials sufficiently close to equilibrium, the reaction rate is under kinetic control. Under kinetic control, the rate of electron transfer at the interface is determined by the electrode potential and is not limited by the rate at which reactant is delivered to the electrode surface (the reaction site). When the electrode potential is sufficiently different from equilibrium, the reaction rate is under mass transport control. In this case, all reactant that is delivered to the surface reacts immediately, and the reaction rate is limited by the rate of delivery of reactant to the electrode surface" [46]. "Faradaic reactions are divided into reversible and irreversible reactions. The degree of reversibility depends on the relative rates of kinetics (electron transfer at the interface) and mass transport. A Faradaic reaction with very fast kinetics relative to the rate of mass transport is reversible. With fast kinetics, large currents occur with small potential excursions away from equilibrium. Since the electrochemical product does not move away from the surface extremely fast (relative to the kinetic rate), there is an effective storage of charge near the electrode surface, and if the direction of current is reversed then some product that has been recently formed may be reversed back into its initial (reactant) form" [46].

"In a Faradaic reaction with slow kinetics, large potential excursions away from equilibrium are required for significant currents to flow. In such a reaction, the potential must be forced very far from equilibrium before the mass transport rate limits the net reaction rate. In the lengthy time frame imposed by the slow electrontransfer kinetics, chemical reactant is able to diffuse to the surface to support the kinetic rate, and product diffuses away quickly relative to the kinetic rate. Because the product diffuses away, there is no effective storage of charge near the electrode surface, in contrast to reversible reactions. If the direction of current is reversed, product will not be reversed back into its initial (reactant) form, since it has diffused away within the slow time frame of the reaction kinetics. Irreversible products may include species that are soluble in the electrolyte, precipitate in the electrolyte, or evolve as a gas. Irreversible Faradaic reactions result in a net change in the chemical environment, potentially creating chemical species that are damaging to tissue or the electrode. Thus, as a general principle, an objective of electrical stimulation design is to avoid irreversible Faradaic reactions" [46].

In certain Faradaic reactions, the product remains bound to the electrode surface. Examples include hydrogen atom plating on platinum and oxide formation. These can be considered a logical extreme of slow mass transport. Since the product remains next to the electrode, such reactions are a basis for reversible charge injection.

2.2. PEDOT Electrodes

Microelectrodes are required for high selectivity neurostimulation. However, the smaller the area is, the higher the impedance. Thus, if the impedance is too high, large voltages will be needed to drive appropriate current. The use of higher voltages can result in the water window overstepping of metal electrodes, which means irreversible Faradaic reactions that can be not tolerated by the physiological system, or can produce electrode corrosion. Besides bigger transistor technologies are necessary in order to deliver higher voltages, it means more surface area for the whole electronic circuit.

The charge injection required for neurostimulation applications is high, and sometimes is difficult to reach within the safe levels of metal electrodes. Due to low charge injection limits, bare noble metals are often disregarded as small-area neural stimulation electrode materials [47] [48].

The impedance of the electrode-electrolyte interface can be drastically decreased by applying a layer of conductive polymer. The decrease of impedance in metal electrodes covered with polymers is mainly related to the increase of Electro-chemical/active Surface Area (ESA) because of the roughness and the porousness of the conductive polymers. Electrodes covered with conductive polymers exhibit high charge delivery capacities due to high ionic conductivity and a large ESA.

Moreover, conductive polymers can be seeded with agents aimed at promoting neural growth toward the electrode sites or minimizing the inherent immune response, and also maintain its electrochemical properties [49] [50].

In [51] a long-term characterization of 16 different electrode materials was performed. The conductive polymers used were Poly(3,4-Ethylenedioxythiophene) (PEDOT) and PolyPyrrole (PPy). In relation to impedance the best results were obtained with the electrodes coated with a layer of conductive polymer. However, the PEDOT showed a better long-term stability than PPy, the impedance of the former remained almost constant while the latter showed an increase in time.

A comparison between PEDOT and PPy under several parameters of deposition was performed in [52]. As results is shown the improvement of electrochemical properties of polymer electrodes vs. uncoated electrodes. The electrode impedance could be reduced by up to two orders of magnitude. The charge injection capacity was about 13 times higher for the coated electrodes.

PPy is unstable and susceptible to degradation under physiological relevant conditions [53]. PPy does not seem to be suitable for long term use in invasive electrodes. Despite of it has good properties like a high conductivity, good biocompatibility and controllable surface properties, its electroactivity is not stable long term [54].

PEDOT shows good electrochemical properties, lower impedance than uncoated electrodes, better charge injection capability, the ability to improve the cell interaction by attaching some agents, and good stability. It can be also deposited in living neural tissue [55]. Therefore, PEDOT is an attractive candidate for invasive stimulation microelectrodes [56] [57] [58] [59].

In the following sections is presented: firstly, the polymer deposition methods; afterwards, our polymerization results on macro and micro electrodes; then, a study of the parameters effect on PEDOT electrical properties.

2.2.1. Selection of Polymer Deposition Method

PEDOT-PSS is industrially synthesized from the EDOT monomer, and PSS as a template polymer using sodium peroxodisulfate as the oxidizing agent. This affords PEDOT in its highly conducting, cationic form. The role of PSS, which has a much higher molecular weight, is to act as the counter ion and to keep the PEDOT chain segments dispersed in the aqueous medium. In general PEDOT-PSS gel particles are formed that possess excellent processing characteristics to make thin, transparent, conducting films. Aqueous dispersions of PEDOT-PSS are commercially available. With this material, thin, highly transparent and conductive surface coatings can be prepared by spin-coating on almost any surface [60].

In the spin coating process an excess amount of polymer solution is dropped on top of a substrate. The substrate is then rotated at high speed at an angular velocity, in order to spread the fluid by centrifugal force, reducing fluid thickness. Rotation is continued for some time, with fluid being spun of the edges of the substrate, until the desired film thickness is achieved, Figure 2.2.1. The solvent is usually volatile, providing its simultaneous evaporation.

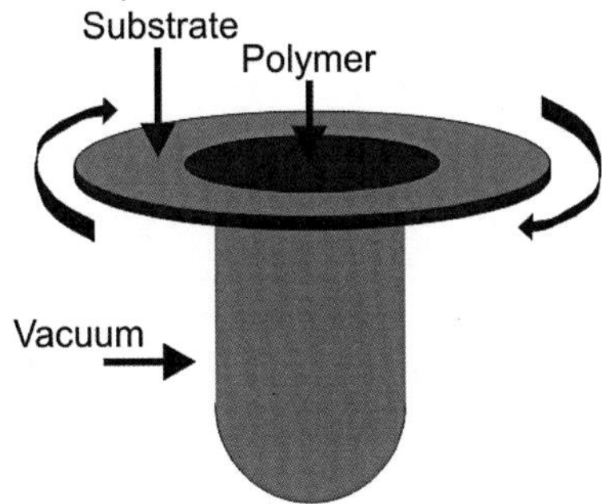

Figure 2.2.1 Polymer spin coating.

We performed some spin coating tests, with them we were looking for the appropriate deposition parameters according our application. With help of the Institute of Optical and Electronic Materials we developed a glass substrate with two parallel gold stripes, the gold deposition was done in an ultra-high vacuum chamber, then we deposited a 100 nm gold layer by electron beam evaporation of a high purity metal in a background of $6.66x10^{-4}$ Pa, the deposition rate was on the order of 3 Å/s. Afterwards, we acquired a high conductivity aqueous dispersion of PEDOT-PSS (Clevios PH 1000, H.C. Starck). We deposited a layer of PEDOT by spin coating process, the parameters used were: 1200 rpm for 30 seconds, according the fabricant the thickness of the layer should be about 100 nm. After that, we annealed the substrate at 150° C for two minutes in order to increase the conductivity of the sample. On each gold stripe we let an uncovered small pad for used it as contact. We tried to measure the conductivity between the gold structures, which must be the conductivity of the deposited PEDOT. We measured some samples, but we had inconsistencies in the measurements, before of reaching quantitative results we found information published by *Nardes* [60], which make us to look for a different deposition method.

Nardes [60] performed dc conductivity measurements of PEDOT-PSS deposited by spin coating process. The tests measured the temperature dependence of the polymer by applying an electrical field perpendicular to the substrate and also coplanar, i.e. parallel to the substrate. Differences between the perpendicular and lateral conductivity of up to three orders of magnitude were found, see Figure 2.2.2. This shows that the spin coated material is highly anisotropic.

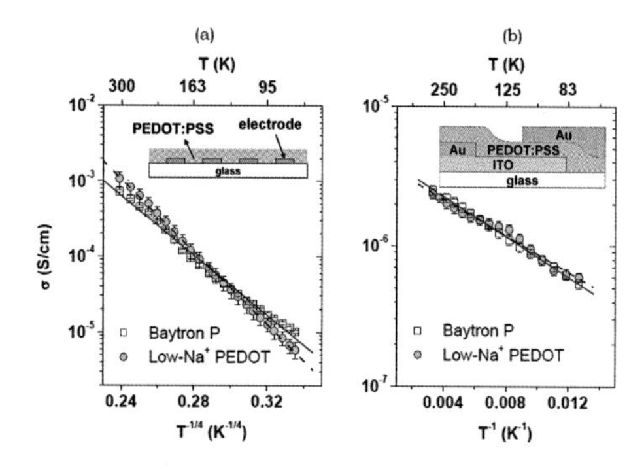

Figure 2.2.2 Conductivity of PEDOT-PSS deposited by spin-coating, a) lateral measurement, b) perpendicular measurement, reprinted from [60].

The physical reason for this behavior is the microscopic morphology of the film. The films consist of flattened PEDOT-rich particles embedded in a PSS-rich matrix of low conductivity. Figure 2.2.3 shows a cross-sectional view of the polymer thin layer. PEDOT-rich clusters (dark) present relatively high conductivity and are organized in layers that are separated by quasi-continuous nm-thick PSS lamellas (bright). PSS shows only ionic conductance, thus, it presents relatively low conductivity. In the perpendicular direction, the PSS lamellas enforce nearest-neighbor hopping between the quasi-metallic PSS particles, leading to a strongly reduced conductivity. The sample is assumed to be inhomogeneous in the vertical direction [60] [61].

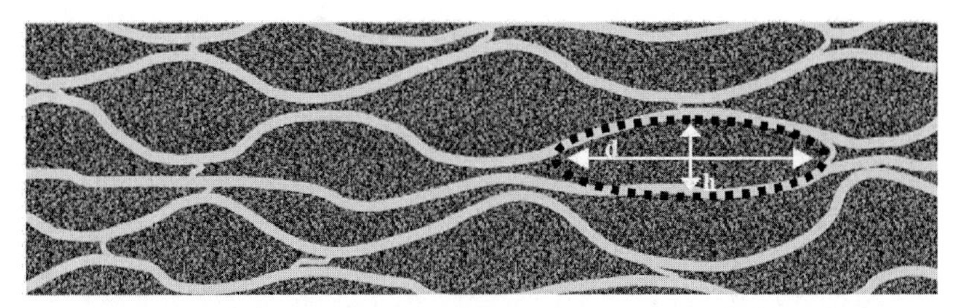

Figure 2.2.3 Cross sectional view of the morphological model for PEDOT-PSS thin film, the typical diameter d of the particles is about 20-25 nm and the height h is about 5-6 nm, reprinted from [60].

Besides, in order to develop our electrodes PEDOT patterning would be necessary. Patterning of PEDOT-PSS is not straightforward because PEDOT-PSS films are damaged by aqueous solutions, which are standard developers in conventional photolithography; and acid-sensitive photoresists are adversely affected by the acidic PEDOT-PSS [62]. The company Orthogonal Inc. has licensed the patent from [62] to make and sell the compatible photoresist.

Because of its morphology, its electrical anisotropic conductivity and the necessity for special compatible lithographic solvents, the PEDOT films deposited by spin coating seem to be inadequate for neural stimulation electrodes. However, the PEDOT can be polymerized directly on the electrodes by electrochemical polymerization.

Electropolymerization is similar to electroplating of metal films. The polymerization occurs by electrochemical reactions. When a potential is applied to the electrode electrons pass to the electrode–electrolyte interface, as shown in Figure 2.2.4. A film of conducting polymer is generated on the electrode in the case of an insoluble polymer.The electro chemical reactions are classifiable into oxidative or reductive reactions. Electrochemical polymerizations can be classified in a similar way. Generally, electropolymerized polymers are synthesized by electrooxidation [63].

Electrochemical deposition of polymer coatings uses electrolytic solutions, which should contain monomeric or polymeric starting materials. By using electrochemical processes, physical or chemical transformation of the dissolved material causes the formation of a solid thick film on the electrode [64].

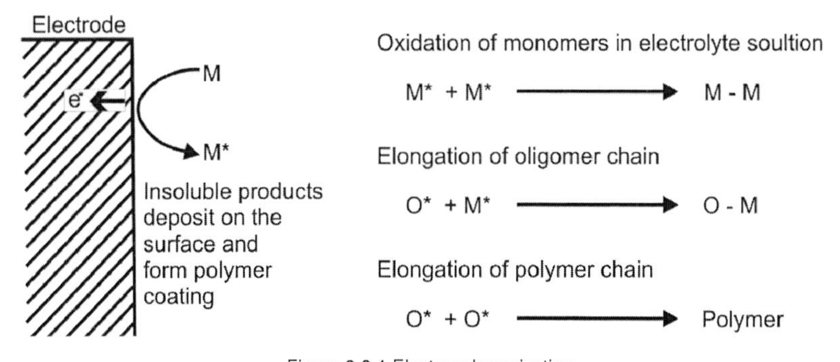

Figure 2.2.4 Electropolymerization.

"Electrochemical polymerization presents advantages. The technique is established and simple. It allows accurate controls over thickness and morphology of the resulting polymer coating on the target electrode surface. Very thin and well-defined coatings can be deposited over microelectrodes in a neural probe or microelectrode arrays. In addition to lateral growth, vertical growth of polymer coatings provides flexibility in the three-dimensional design of electrodes" [65].

"Electrochemical polymerization can be carried out by constant current (galvanostatic method) or constant voltage (potentiostatic method). Although both methods produced quality coatings, the galvanostatic method appears easier to control. When a constant current is applied to the electrochemical cell in a potentiometric mode, the charge used for polymer deposition is simply determined by time and the film grows at a steady rate. However, when a constant voltage is used in a potentiostatic method, the resulting electrode current varies and depends on many factors including the base electrode material, polymer systems, and plating conditions" [65]. The charge calculation for the deposition is not simple and the film growth is less stable.

The morphology of a Polybithiophene (PBT) film is affected by the current density used during the electrochemical polymerization. Current densities between 0.5 and 2 mA/cm^2 create uniformly structured PBT layers with rough surfaces. Current densities of 0.5 mA/cm^2 create smother films. Current densities larger than 2 mA/cm^2 reduce the adhesion of the PBT layer on the electrode [66].

The charge density is associated to the quantity of material deposited on the electrode.

2.2.2. PEDOT Electrodes Results

Results with Pt wires as Macroelectrodes

An aqueous monomers solution was prepared in order to perform the electrochemical polymerization on the electrodes. The solution was prepared in deionized water with a monomer EDOT concentration of 0.01 M, and a surfactant NaPSS concentration of 0.10 M. The PSS powder was mixed into the deionized water then stirred for a couple of minutes until no powder residue was seen in the glass container. After that the EDOT solution was spilled into the deionized water and stirred. All procedures were made at room temperature, about 24° C.

The solution mentioned above was used to perform electropolymerization, and deposit a thin layer of PEDOT on three platinum wires. One wire at a time was submerged into the solution then a specific charge was transferred to the wire, the current and time were controlled. The pulse is driven all over the platinum wire while the counter electrode was a platinum foil, see Figure 2.2.5.

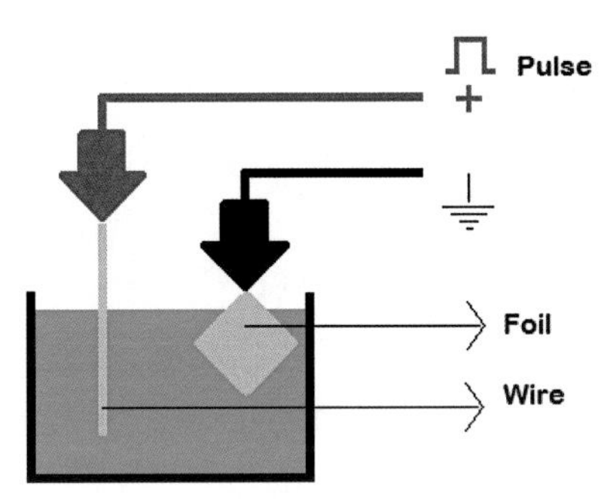

Figure 2.2.5 Pt wires electropolymerization.

The wire diameter was 75 μm and just 1 cm was submerged, the resulting area was 2.36 mm^2, the platinum foil had a dimensions of 25 x 25 mm^2, resulting in an area of 625 mm^2. Both of them where previously cleaned with acetone, isopropanol and water. The current applied had amplitude of 15 μA, for 47, 94 and 188 seconds, respectively to each wire. Resulting in a current density of 0.64 mA/cm^2, and a total charge of 30, 60 and 120mC/cm^2. The PEDOT deposition was assumed due to the blue color on the wires.

The impedance of the wire before and after the deposition was measured by using an Agilent 4284A LCR-meter, with a two electrodes configuration, against the same foil. The measurements were done in saline solution. The frequency was swept from 20 Hz until 100 kHz. The voltage applied was 50 mV, this value was used as standard in all the measurements. With this value, we avoided the non-linear impedance behavior of the Pt [67]. And since this is the common value used, we can compare the results against other publications.

The results of the wires impedances and phases are plotted in Figure 2.2.6. The red dashed line corresponds to the values of the platinum wire, this is just a bare platinum wire; the resting lines correspond to the PEDOT wires with different charges.

It can be seen how for higher charges the impedance decreases, this is related to the thickness of the PEDOT layer. By taking the results of the PEDOT deposited with a charge of 120 mC/cm^2, the impedance of the PEDOT wire against the bare wire was around 96 times lower at 20 Hz, and 6 times lower at 1 kHz.

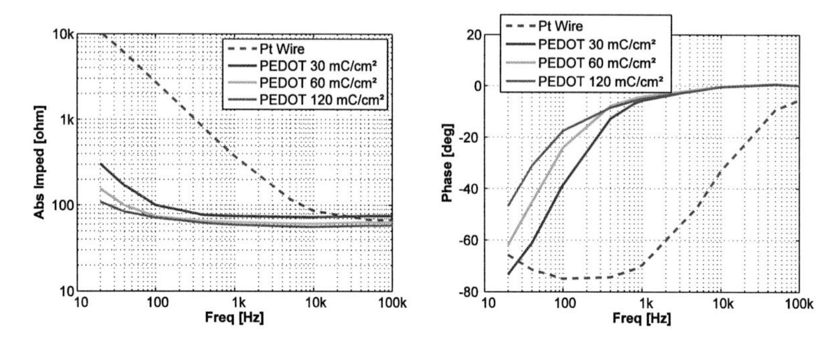

Figure 2.2.6 Impedance and phase on Pt and PEDOT wires with different charges.

Result with Microelectrodes

An experiment was performed in order to corroborate the impedance decrease of PEDOT layers in microelectrodes. The monomer solution preparation was the same as in the previous section. The electrode used was a 30 µm diameter gold electrode on a silicon substrate. The electrode form part of an electrode array, its fabrication is described in Appendix A. The electropolymerization parameters were: a current of 3.5 nA for 120 seconds, resulting in a current density of 0.5 mA/cm² and a total charge of 60 mC/cm². An image of electrode before and after de deposition is shown in Figure 2.2.7.

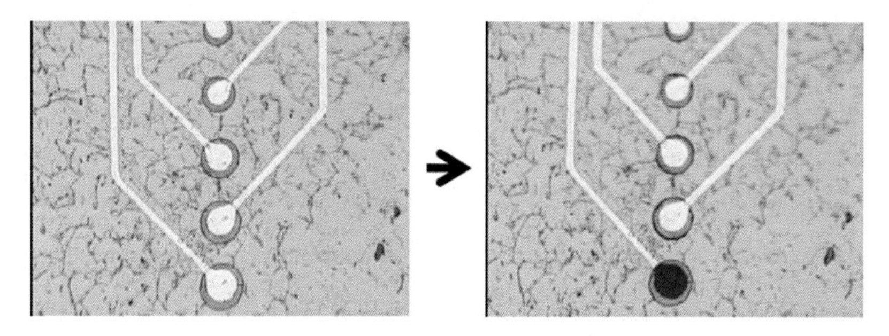

Figure 2.2.7 Polymer deposition on micro electrodes in a silicon substrate.

The resulting impedance and phase are shown in Figure 2.2.8. The impedance of the PEDOT compare to the gold electrode is 160 times lower at 20 Hz, and 25 times lower at 1 kHz.

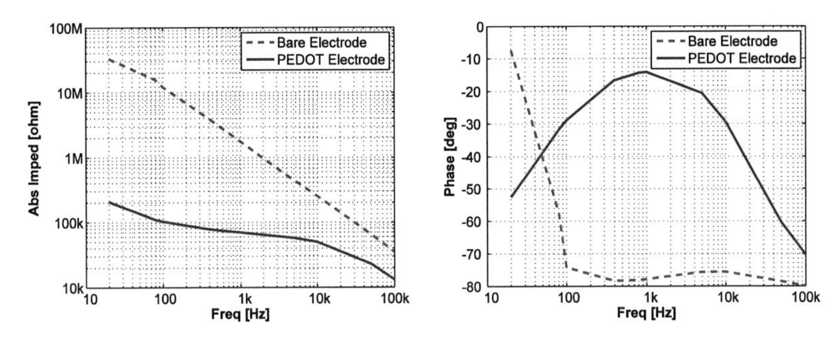

Figure 2.2.8 Impedance and phase on micro electrodes in a silicon substrate.

In vivo Results

An animal experiment was conducted to corroborate the stability and impedance measurements between *In Vitro* (saline solution) and *In Vivo* experiments. The experiment was performed by physicians at the UKE (Universitäts Klinikum Eppendorf) and with the corresponding permissions using a 100 kg pig. The electrodes were inserted between the muscle and skin and the impedance of the gold and PEDOT electrodes was measured.

For the experiment we fabricated a gold multiarray electrodes on a flexible substrate, its area was 1 mm², the fabrication corresponds to the electrodes version 1 described in Appendix B. The electropolymerization parameters were: a current of 5 µA for 120 seconds, resulting in a current density of 0.5 mA/cm² and a total charge of 60 mC/cm². The impedance measurements were done between two electrodes of the multiarray, which have an interdistance of 1 cm.

The results are shown in Figure 2.2.9. The tendency of impedance decrease with PEDOT is also visible. However, there is a difference of about 2 times between the measurement *In Vitro* and *In Vivo*, especially at higher frequencies.

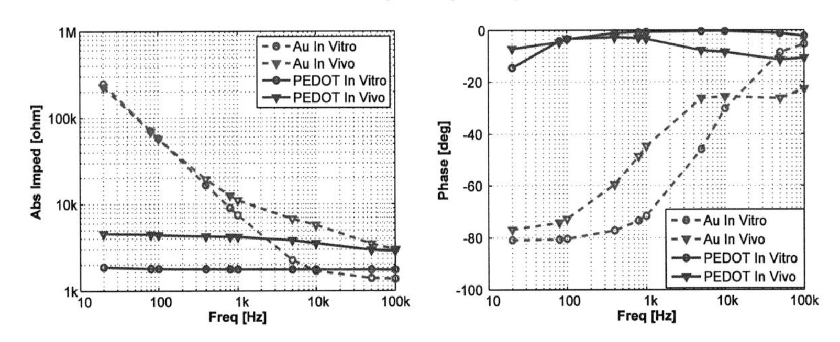

Figure 2.2.9 *In Vivo* Impedance and phase on electrodes in a flexible substrate.

The difference is attributed to the temperature, presence of organic species, uncertain concentration of electrolytes in biological tissues, among other factors [45]. The *In Vitro* experiments were performed at room temperature, about 24° C. For the *In Vivo* experiments a pig corporal temperature of 38.8° C is expected. The impedance increase behavior because of the temperature is explained in Equation 2.8, Equation 2.9 and Equation 2.10.

2.2.3. Effect of the Electropolymerization Parameters

In this section is shown the influence of the electropolymerization parameters on the electrical properties of PEDOT layers over gold electrodes. We performed this study in collaboration with the Institute of Optical and Electronic Materials, some results were published on [68] and are part of the dissertation [69].

The chemicals used were the followings: Monomer 3,4-ethylenedioxythiophene (EDOT) and surfactant sodium polystyrenesulfonate (NaPSS) (MW=70000 g/mol) were used without further purification. An irrigation solution (saline solution) was used to perform the *In Vitro* measurements.

For the experiment we fabricated a gold multiarray electrodes on a flexible substrate, the fabrication corresponds to the electrodes version 2 described in Appendix B. Following deposition, a positive photo-resist (S-1805) was used to pattern specific macroelectrodes (0.5 mm^2, 1.0 mm^2 and 2.0 mm^2) and microelectrodes (177 μm^2, 413 μm^2 and 1000μm^2).

PEDOT/NaPSS was electropolymerized from 10 mM EDOT in 2.0 g/100mL NaPSS aqueous dispersion under galvanostatic conditions. The current densities were swept from 0.1 to 10 mA/cm^2 with varying charge density (40, 80, 120, 160, 240, 300 mC/cm^2). The current output was controlled by a Keithley 224 Programmable Current Source in a two-electrode cell configuration with a platinum foil as the counter electrode.

The samples were characterized by an Agilent 4284A LCR-meter. Impedance measurements were executed with two electrodes set up from 20 Hz to 1 MHz with excitation amplitudes of 50 mV in NaCl-KCl saline solution (7.4 pH). Pulsing was performed in non-agitated solution against a large area of stainless steel counter electrode. Cyclic Voltammetry (CV) measurements of PEDOT were evaluated by Wenking STP84 potentiostat from -800 to +600 mV in a slow sweep rate (50 mV/s). A platinum foil served as counter electrode and an Ag|AgCl|KCl (3M) was used as the reference in a NaCl-KCl saline solution (7.4 pH). All data were recorded by a Keithley 4200-SCS machine.

CV measurements were used to evaluate the charge injection (Q_{inj}) capability of the PEDOT/NaPSS films. Q_{inj} was calculated by integrating the cathodal current density enclosed by the CV and divided by the sweep rate.

The current density effect was studied, Figure 2.2.10. PEDOT films were synthesized galvanostatically sweeping from 0.1 to 10 mA/cm^2 and with a fixed charge density of 80 mc/cm^2 in 1 mm^2 macroelectrodes. The data showed no significant effect on the impedance values at relatively low current densities (<2.0 mA/cm^2), but low currents are recommended to create more uniform and structured layers [66] [70]. Above 2.0 mA/cm^2, high impedance values were detected, together with substrate (gold) delamination. The following tests were performed with a fixed current density of 0.5 mA/cm^2.

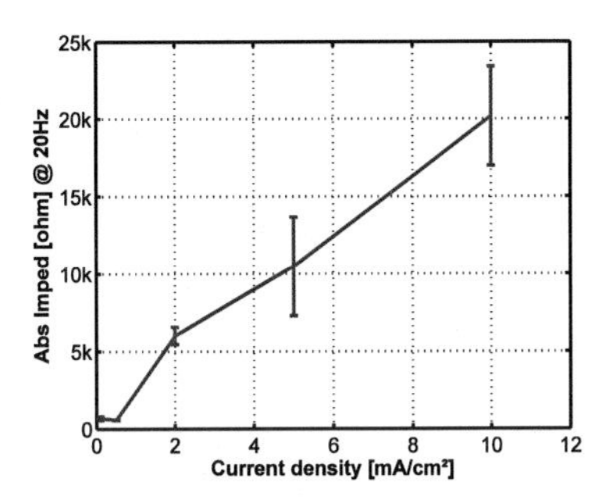

Figure 2.2.10 Current density effect on PEDOT electropolymerizaton.

For PEDOT coats some difference of the impedance tendency was observed for macro and microelectrodes at 20 Hz in function of charge density used during the electrodeposition. The Figure 2.2.11 shows how for microelectrodes (1000 μm²) there is a significant decrease in impedance up to 160 mC/cm², from this point the impedance decreases slowly. While for macroelectrodes (1.0 mm²) the lowest impedance value is reached at 80 mC/cm². This variation can be attributed to the lack of monomer in the negative microcavity while the polymerization goes on. It is known that a low monomer concentration reduces the adhesion and leads to less uniform polymer layer [70].

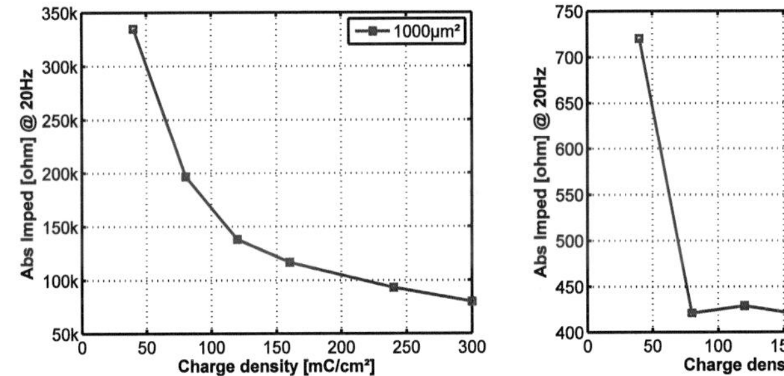

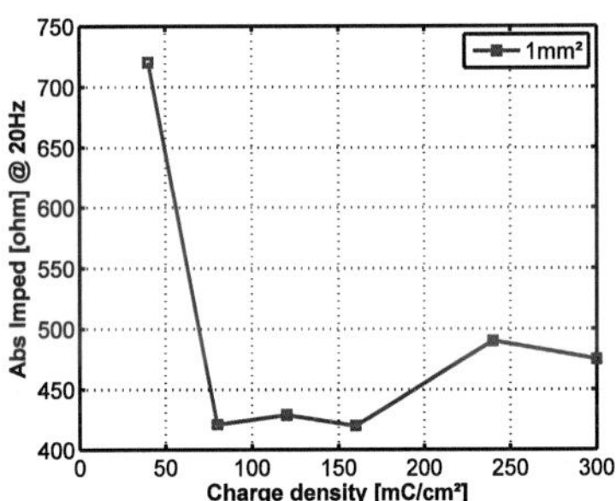

Figure 2.2.11 Charge density effect on PEDOT electropolymerizaton.

In addition, the charge injection (Q_{inj}) capability of the PEDOT coat showed linear relationship with the charge density used during the polymerization, Figure 2.2.12. This phenomenon is associated with a highly effective surface area, Faradic reactions and high capacitance constant values in the polymer layer [56] [71]. Zero charge density correspond to the bare electrodes, i.e. gold electrodes.

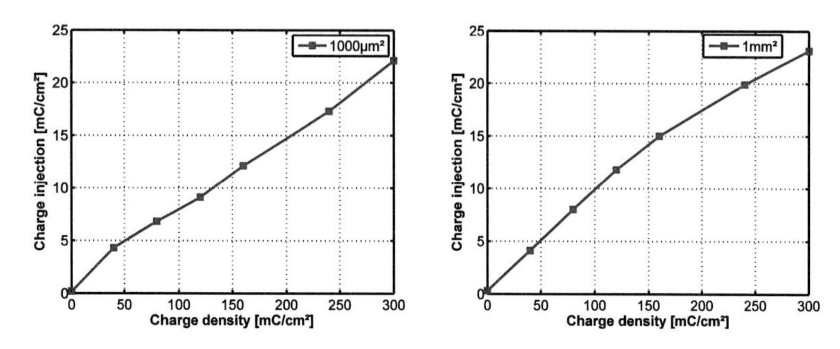

Figure 2.2.12 Charge injection effect on PEDOT electropolymerizaton.

The transferred charge is found to be proportional to the thickness of the resulting layer, Figure 2.2.13. This phenomenon is present because the charge density is associated to the quantity of material deposited on the electrode. Same tendency is observed for the charge injection which confirms the increase of Faradaic phenomenon when the quantity of material increases.

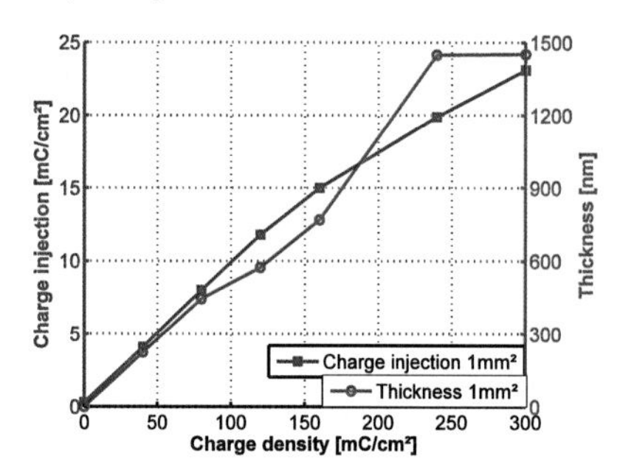

Figure 2.2.13 Thickness effect on PEDOT electropolymerizaton.

With more PEDOT deposition, the charge injection increases proportionally and the impedance still decreasing. However, thick PEDOT layers seem to be delaminated under prolonged stimulation [58]. Cui and Zhou tested PEDOT electrodes with pulsing currents for a period of two weeks, the layers deposited with around 112 mC/cm^2 showed cracks and the layers deposited with around 230 mC/cm^2 presented delamination after some days.

Figure 2.2.14 shows an inverse relation between the impedance and the electrode area at 20 Hz. Impedance at high frequency range depends predominantly on solution resistance and the impedance in this range is not inversely proportional to the electrode area, since at high frequency values, capacitive components of electrolytes and double layer capacitance of the electrode surface have negligible effects on the impedance [54].

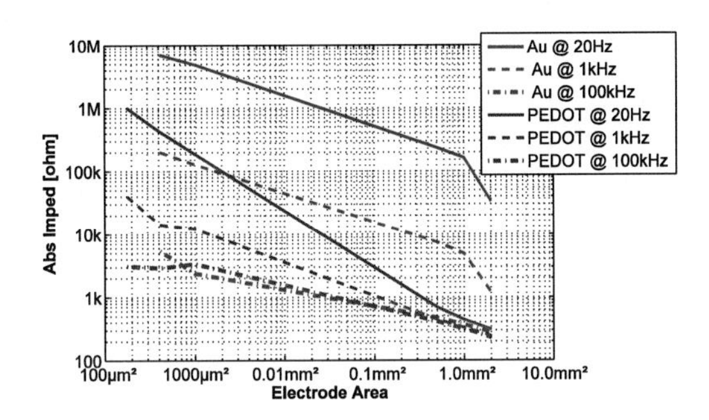

Figure 2.2.14 Size effect on PEDOT electropolymerizaton.

In Figure 2.2.15 and Figure 2.2.16, the impedance and phase of micro and macroelectrodes are shown, respectively. They were deposited with a charge density of 80 mC/cm².

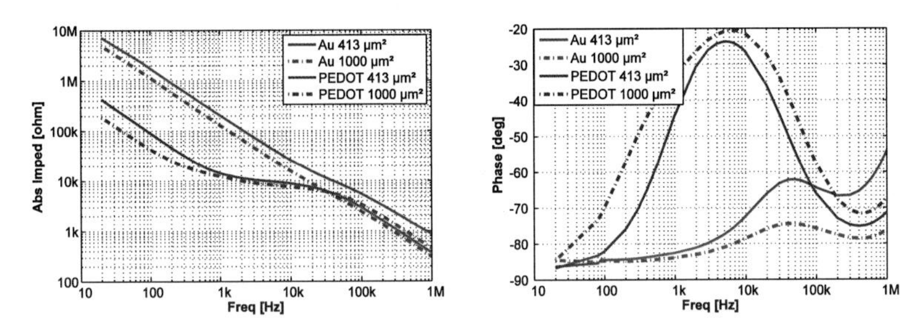

Figure 2.2.15 Impedance and phase of microelectrodes.

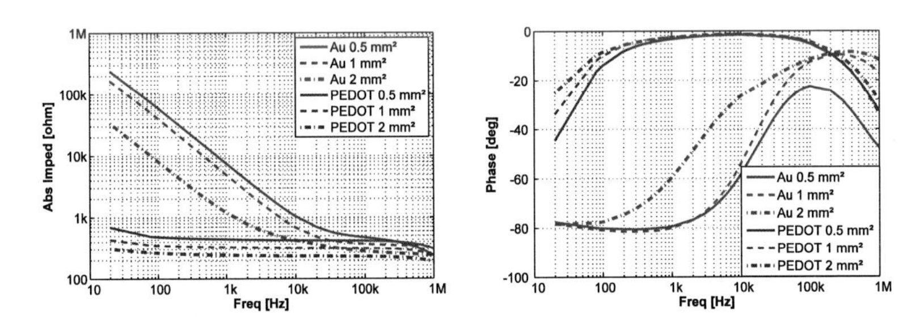

Figure 2.2.16 Impedance and phase of macroelectrodes.

2.3. Chapter Conclusion

Faradaic electrodes provide more charge per unit geometric surface area (GSA) than do capacitive electrodes. Faradaic charge injection capacity may also be increased by increasing electrochemical surface area. The electrochemical surface area (ESA) can be increased by roughening or by deposition of highly porous electrode coatings, such as PEDOT.

PEDOT shows good electrochemical properties, lower impedance than uncoated electrodes, better charge injection capability, the ability to improve the cell interaction by attaching some agents, and good stability. PEDOT is therefore an attractive candidate for invasive stimulation microelectrodes.

PEDOT could be deposited using different methods, such as spin coating and electrochemical polymerization. However, because of its morphology, its electrical anisotropic conductivity and the necessity for special compatible lithographic solvents, the PEDOT films deposited by spin coating seem to be inadequate for neural stimulation electrodes. Thus, PEDOT can be polymerized directly on the electrodes by electrochemical polymerization.

The electrochemical deposition is a simple method to deposit the PEDOT layer on the electrodes, and it is possible even when the implants are already assembled so long as the stimulator circuitry is able to deliver the current under the necessary parameters to perform the electropolymerization.

Regarding the polymerization parameters, low current densities are recommended to avoid substrate degradation and to improve the polymer package. The transferred charge is found to be proportional to the thickness of the resulting layer. Thicker layers improve the Q_{inj}, and reduce the impedance in the interface. However, it is necessary to find the optimal point where Q_{inj} and impedance are improved and the layer does not show delamination.

PEDOT/NaPSS electrical properties (Impedance and Q_{inj}) were constant and reproducible after several experiments, even *In Vivo*. These parameters can be estimated for macro and micro electrodes in function of the charge density.

The PEDOT layers galvanostatically deposited show good adhesion in Pt and Au, materials which present biocompatibility and good tissue response.

3. Waveform for Stimulation

The rectangular pulse waveform has been used as standard shape for electrical neurostimulation. However, through simulations, it has been shown that non-rectangular waveforms can provide more energy-efficient neural stimulation and also reduce stimulation artifacts [72] [73].

While the strength-duration curve is defined for rectangular pulses, different pulse shapes shift the chronaxie time. Thus, by injecting longer pulses, it is possible to improve the charge injection of the electrodes and also to reduce the threshold charge and threshold energy [74].

The results of some studies suggest that non-rectangular waveforms are more comfortable to the patient in Neuromuscular Electrical Stimulation (NMES) [75].

It is desirable to stimulate small groups of neurons or even single neurons. Thus, it is convenient to build small electrodes in order to achieve high selectivity. However, with smaller electrodes, higher stimulation voltages are required, due to the electrode impedance. Through current stimulation with different waveforms it is possible to reduce the voltage peak of the injected signal necessary to achieve the firing of an action potential [76].

It was also demonstrated that the waveform and frequency of the signal have certain influence on the selectivity of stimulation of neurons with their cell bodies near the electrode and fibers of passage [77][78].

In order to understand the behavior of the stimulation with different waveforms, we performed simulations by using a passive mammal membrane model, section 3.1. Afterwards, in order to verify whether the same results still present by transcutaneous stimulation, we develop an NMES equivalent electrical circuit, which we used to perform simulations with different waveforms, section 3.2. The results showed that non-rectangular waveforms can provide more energy-efficient neural stimulation. Therefore, we developed a portable stimulator, section 3.3, which enables experimentation with several stimulation parameters. Afterwards, by using our stimulator, we performed *In Vivo* tests to corroborate the simulations, section 3.4. The experiments were done by applying transcutaneous electrical muscle stimulation to different subjects.

3.1. Simulations with a Passive Membrane Model

In order to understand the behavior of the stimulation with different waveforms, we perfomed simulations by using a mammal membrane model. The Hodgkin-Huxley model, shown in section 1.1.9, describes the behavior of the channels in the membrane and its voltage and time dependence. That model fits to describe how action potentials in neurons are propagated. However, the scope of this simulation is to analyze under which circumstances the firing of an action potential is achieved. That can be observed by using a passive model from [79], which used a fix value for the resting membrane potential (V_r), the membrane conductance (G_m) is the lumped effect of all the ionic channels and the double-layer lipid, which separates the internal and the external charges is modeled as a capacitance, C_m. The equivalent electrical circuit is shown in Figure 3.1.1.

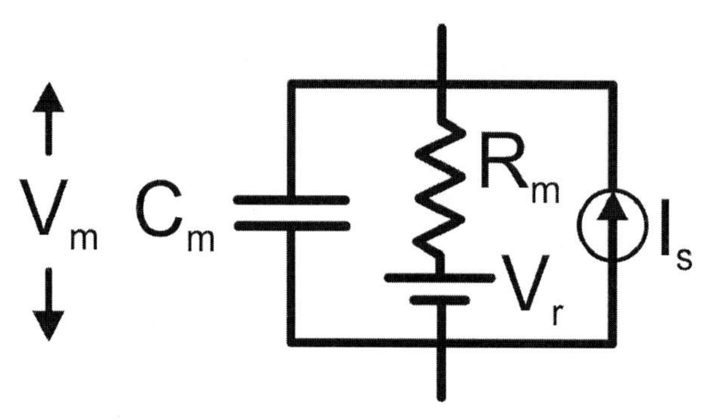

Figure 3.1.1 Passive membrane model.

Where, R_m is the modeled electrical resistance, the inverse of G_m, which is usually given by the multiplication of the membrane area with the unit conductance g_m. The membrane capacitance C_m can be also calculated by multiplying the capacitance of unit membrane area, c_m (F/cm^2), with the total membrane area. And I_S is the current externally applied to the membrane. In the Figure 3.1.2, the calculation of the membrane area is depicted.

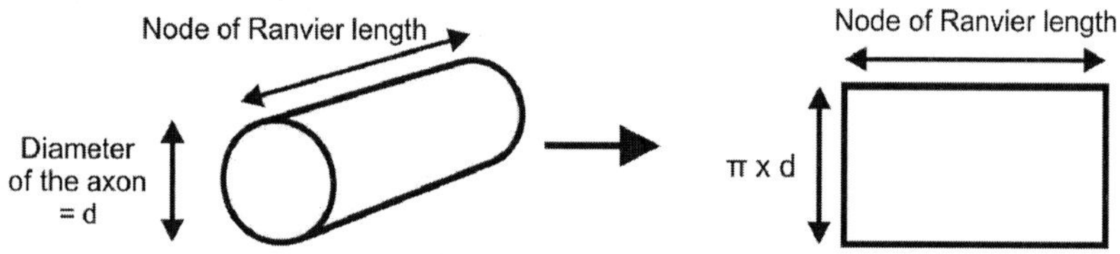

Figure 3.1.2 Membrane area calculation.

The dimensions of the axon are taken from the Table 1.1.1, the node of Ranvier length is 4 µm and the diameter 20 µm. The membrane resting potential, V_r, according the section 1.1.5 is usually about – 70 mV. The values of the unit conductance, and unit membrane capacitance are taken from [74], the resulting values are shown in Table 3.1.1.

Table 3.1.1 Values of the membrane used for the passive model.

	Unit value [74]	**Resulting value**
C_m	2.5 µF cm^{-2}	6.28 pF
G_m	128 mmho cm^{-2}	0.32 µmho
V_r	N/A	- 70 mV

We want to observe the membrane voltage in function of the time, the equation for the voltage on a capacitor is:

Equation 3.1 $$V_c(t) = \frac{Q(t)}{C_m} = \frac{1}{C_m}\int_{t0}^{t} I_C(t)dt + V_c(t0) = V_m(t)$$

In order to calculate the voltage on the membrane, the same as the voltage on the capacitor, we need to know the current flowing through the capacitor, $I_C(t)$. By Kirchhoff´s law:

Equation 3.2 $$I_C(t) = I_S(t) - I_R(t)$$

Where, the current flowing through the capacitor is equal to the stimulation current minus the current on the resistor. By Ohm´s law:

Equation 3.3 $$I_R(t) = \frac{V_m(t) - V_r}{R_m} = (V_m(t) - V_r)G_m$$

Then:

Equation 3.4 $$V_m(t) = \frac{1}{C_m}\int_{t(0+1)}^{t}[I_S(t) - (V_m(t-1) - V_r)G_m]dt + V_m(t0)$$

In order to simulate the voltage in Matlab we derivated the last equation by using differential steps, and the resulting expression is:

Equation 3.5 $$V_m(t) = \frac{\left(\frac{I_S(t)}{G_m} + V_r\right)\Delta + V_m(t-1)\tau}{\Delta + \tau}$$

Where $\tau = (C_m/G_m)$ and Δ is the time step, which was set to 0.1 µs. In the Figure 3.1.3, the current stimulus with different waveform is depicted, the membrane voltage for each waveform is also in the figure, the parameters used were 50 µs pulse width and 15 nA current amplitude.

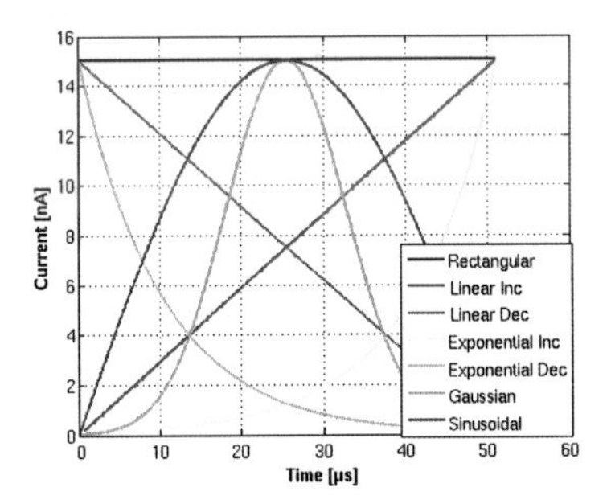

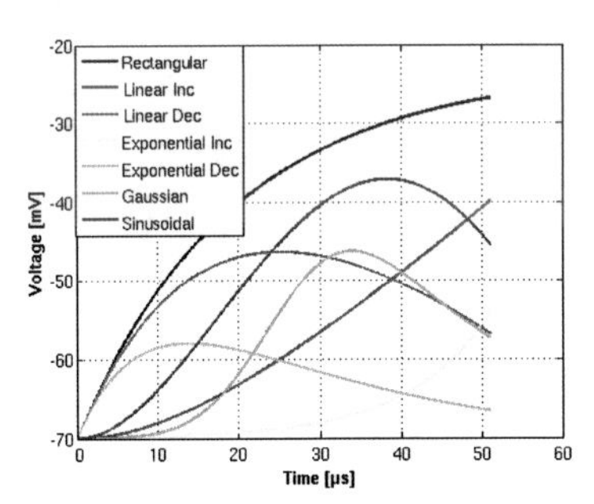

Figure 3.1.3 Current applied and membrane voltage.

In order to calculate the strength-duration curve for the different waveforms, we implemented an algorithm in Matlab, which makes a pulse width sweep, from 1 µs to 350 µs, with a time step of 1 µs. For every pulse width, it tests each waveform from the minimum current amplitude, 0.1 nA, if the membrane voltage threshold is reached, - 50 mV, then continues with the next pulse width, if not, increases the current amplitude in steps of 0.1 nA and tests the waveform again. The resulting strength-duration curve is shown in Figure 3.1.4.

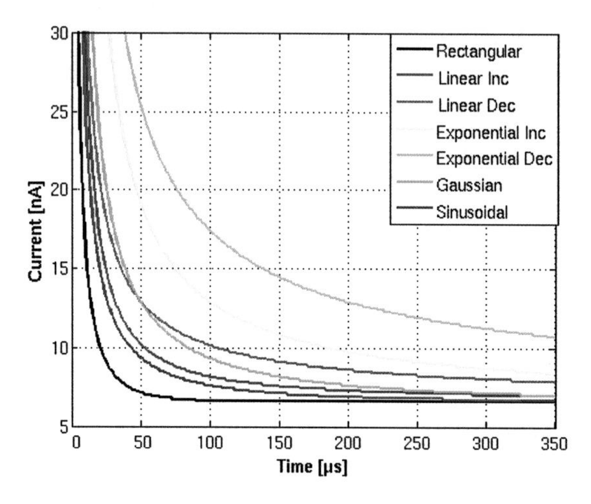

Figure 3.1.4 Current strength-duration curve from the passive membrane model.

The curve shows the same characteristics as the Figure 1.1.7, when the strength-duration curve was introduced. However, the chronaxie for the non-rectangular waveforms is shifted, and is necessary to stimulate with longer pulses and higher current amplitudes in order to achieve the activation of the cell. But, if the goal is find the optimal stimulation waveform, there are different parameters that have to be considered. The total amount of charge transferred to the membrane is shown in the Figure 3.1.5.

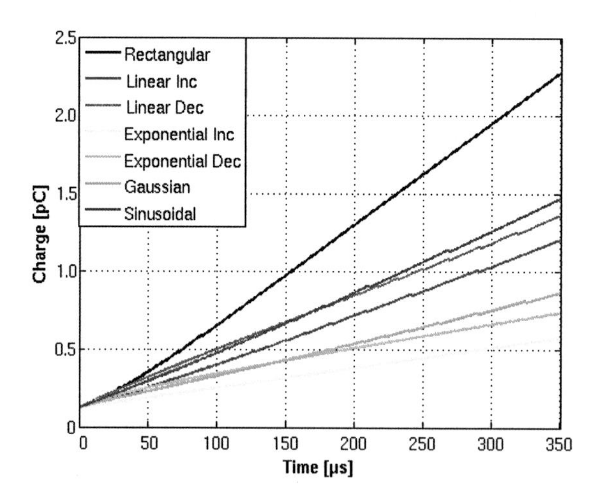

Figure 3.1.5 Charge threshold.

Until 30 µs the necessary charge is very similar with the different waveforms, from this point, by stimulating with rectangular waveforms the required charge is higher.

The Figure 3.1.6 shows the energy threshold, it was calculated by multiplying the current with the voltage over the time.

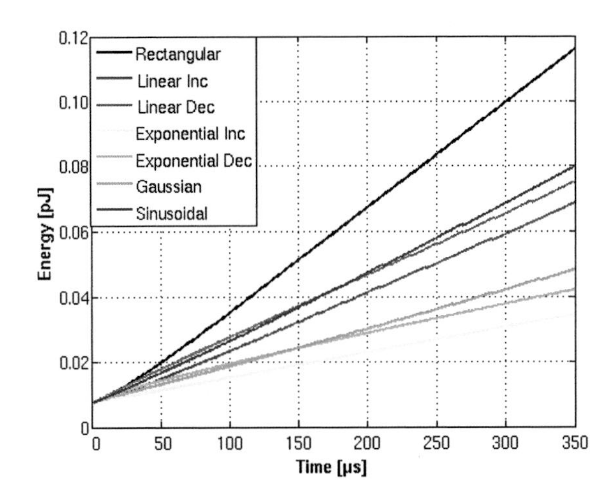

Figure 3.1.6 Energy threshold.

It shows the same behavior as the charge. It is necessary to observe the strength-duration curve, for short pulses a high current amplitude is necessary at the electrodes, which must cause high voltage peaks, and it would increment significantly the energy required. However, the electrode circuit is not implemented in this model, and it lacks of a resistive element in series with the capacitor. Because of it, the last curve is not the most appropriate way to predict the energy necessary at the electrodes. A good approach is the equation used in [74]:

Equation 3.6 $$E_{th}(t) \propto \int_0^t \mathrm{I_{th}(t)^2 dt}$$

There, the energy is just a proportional value because of the absence of electrode impedance. By using the last equation, the resulting curve for the energy threshold is shown in Figure 3.1.7.

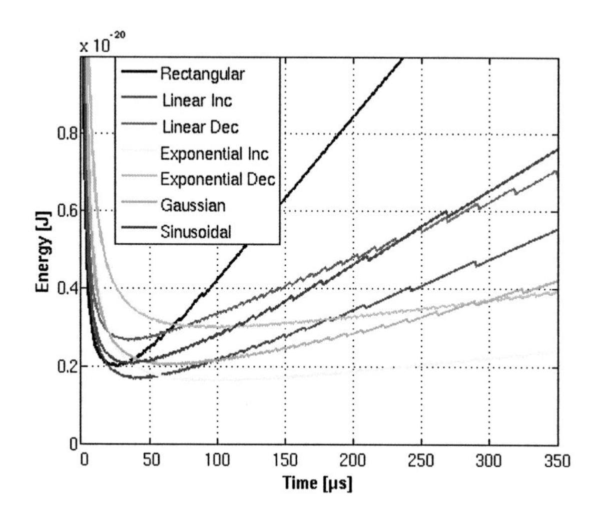

Figure 3.1.7 Energy threshold calculated with the Equation 3.6.

In this curve, it can be seen that the minimum energy threshold is found at its corresponding chronaxie time. The next table summarizes the characteristics of the different waveforms analyzed:

Table 3.1.2 Summarize of the stimulation parameters with different waveforms by using the passive membrane model.

Waveform	Chronaxie [μs]	Current threshold [nA]	Charge threshold [pC]	Energy threshold [J] x 10^{-20}
Rectangular	27	8.70	0.232	0.199
Linear increase	43	10.90	0.232	0.167
Linear decrease	34	15.40	0.260	0.266
Exponential increase	95	13.20	0.246	0.162
Exponential decrease	92	18.10	0.329	0.299
Gaussian	65	11.20	0.258	0.202
Sinusoidal	38	10.50	0.251	0.205

According the last table, by stimulating with non-rectangular waveforms, and at its corresponding chronaxie time, the required energy can be decreased around 20%. However, an electrical model that includes the electrodes and tissues along the path is necessary, in order to obtain an accurate calculation, and voltage levels at the electrodes.

3.2. Simulations with a NMES Equivalent Electrical Circuit

In the last section it was shown that energy can be saved by stimulating with non-rectangular waveforms. However, in order to have more accurate results, the incorporation of equivalent circuits for the electrodes and tissues along the path is necessary. The author of this document proposed and supervised a master thesis, which consisted in the development of a NMES equivalent electrical circuit. The equivalent circuit model was also developed with the aim to study the neuromuscular electrical stimulation and to provide an analysis of the strength-duration curve. In this section the model is presented, its development, and the results of simulation with different waveforms. The master thesis was presented by *Lujan* [80] and further improvements has been published [81] [82].

3.2.1. TENS Stimulation Experiment

The first step to develop the equivalent electrical circuit was to perform electrical neuromuscular stimulation on a subject, the experiment was done by using a Transcutaneous Electrical Nerve Stimulation (TENS) device. The aim of the experiment was to obtain the strength-duration curve, and to record the electrode´s voltage over the time. Then, the measured values were used as reference for the equivalent model.

One male subject participated in the study. Subject's age and mass were 27 years and 68 kg, respectively. Adhesive carbon electrodes with dimensions of 40 mm x 40 mm were utilized in the experiment. The inter-electrode distance was 40 mm. The configuration used was with bipolar electrodes. One electrode was placed on the beginning of the biceps brachii distal of the shoulder. The last electrode was placed 40 mm distal of the first electrode, close to the cubital fossa.

The experiment was performed with a transcutaneous electric stimulator (TENS Plus 690550, Megro GmbH & Co. KG). The device can deliver electric current up to 80 mA with monophasic rectangular pulses from 30 μs to 260 μs. First, the rheobase was determined with a monophasic rectangular pulse with a pulse width of 260 μs and frequency of 2 Hz. The intensity was increased until a barely muscle twitch was visible. The strength-duration curve was determined by decreasing the pulse width (220, 200, 180, 160, 150, 140, 120, 100, 80, 60, 30 μs) and then increasing the electric current until a twitch was visible. Measurement errors were present since the device has not a digital control and the values were set by turning the knob by hand. An oscilloscope was used to measure the voltage at the electrodes, and the current flowing through them, with the use of a resistor in series.

The strength-duration curve was found normal for a flexor muscle in terms of the shape, see Figure 3.2.1. Chronaxie value was around 134 μs.

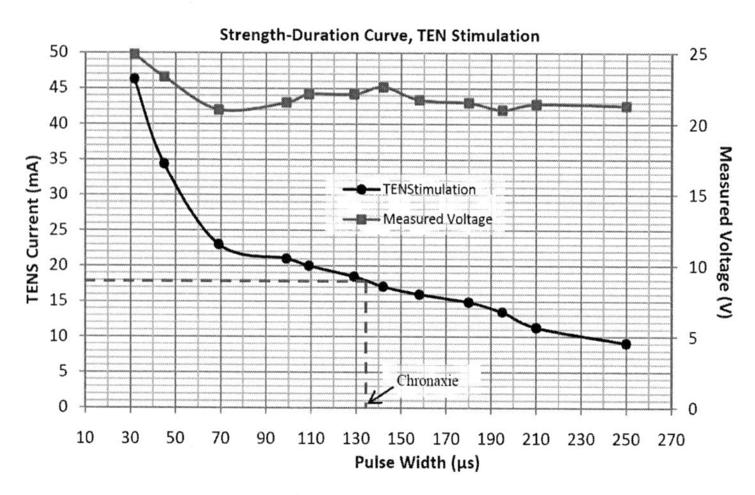

Figure 3.2.1 Strength duration curve and measured voltage outcomes of the transcutaneous electrical nerve stimulation, reprinted from [80].

3.2.2. NMES Equivalent Circuit Model

In this section, the different stages of the model are described, a non-linear element is incorporated and the complete model is presented.

Simple Equivalent Circuit Model

The traditional equivalent circuit of transcutaneous electrode placed on skin is depicted in Figure 3.2.2. The figure illustrates the equivalent circuit for every stage that is taken into consideration.

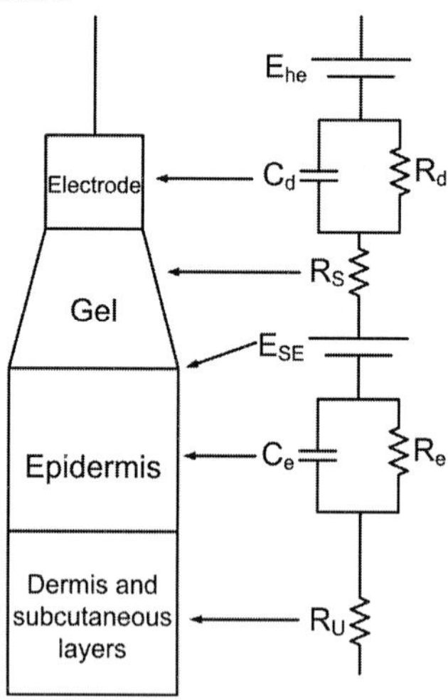

Figure 3.2.2 Traditional equivalent circuit of transcutaneous electrode placed on the skin.

The first equivalent circuit illustrated is the electrode-electrolyte interface. This interface was described in the section 2.1. The half-cell potential (E_{he}) is due to the electric field formed on the interface. The potential for carbon electrodes is between 0.03 V and 2 V [83].

The equivalent circuit for the skin begins with the equivalent circuit of the epidermis. This level has several sub-layers which are neglected in this equivalent circuit; nevertheless, it is represented as an RC circuit representing the total impedance of this tissue. The epidermis is the outer layer of the skin and is composed of several layers of cells. The thickness varies in different sites of the body. On the soles of the feet and the back it is quite thick measuring 1.5 mm, but in other areas of the body, e.g. the eyelids, the epidermis consists only 0.10 mm [84]. After the epidermis equivalent circuit it comes the equivalent circuit of the dermis and subcutaneous fat, which is characterized by only a resistor. In the next section, these layers will be further developed; nevertheless, the resistor represents the total impedance for these layers. Some authors [85] [86] utilized a RC parallel circuit which represents the skin (epidermis, dermis and subcutaneous fat) with a resistor in series characterizing the muscle tissue (or deep tissue).

Equivalent Circuit of the Electrode-Electrolyte Interface

The electrolyte contains free ions that make it electrically conductive. The stratum corneum is composed mainly of dead cells; therefore, it is necessary to notice that the electrolyte in the interface is the conductive gel which is placed between the electrode and the skin.

The equivalent circuit is based on the Randles model showed in 2.1.2. The double layer impedance is viewed as a resistance R_d, a capacitance C_d, and the half cell potential E_{he}. The typical resistance is 10 kΩ, whereas the capacitance is 10 µF [87].

Equivalent Circuit of the Stratum Corneum

The stratum corneum is the outmost layer of the epidermis and also of the skin which consist of dead cells. The thickness of the stratum corneum varies according to the amount of protection required by a region of the body and it is generally between 10 µm and 40 µm. The impedance of the skin is dominated by the stratum corneum at low frequencies and it was stated that at frequencies below 10 kHz the skin impedance is determined by this thin layer. The stratum corneum is a solid state layer, not necessarily containing liquid water and it may be considered as a solid state electrolyte. It includes a few ions free to move and contribute to direct current conductance [88].

Equivalent Circuit of the Epidermis, Dermis and Subcutaneous Fat (lower layers)

Human skin is a very inhomogeneous material and its electric properties differs. The epidermis has five different layers; however the layer that defines the skin impedance is the outmost layer, the stratum corneum. Deeper skin layers named: the rest of epidermis, dermis and subcutaneous fat have much lower resistivity [89].

Homogeneous electric properties in the rest of the epidermis, dermis and subcutaneous fat can be assumed [90]. Thus, we used just one equivalent circuit to represent those layers.

In the Figure 3.2.2, we observed that the equivalent circuit of the dermis and subcutaneous fat was just a resistance. Petrofsky [91] suggests that a RC low pass filter is created by fat. Thus, any model must take subcutaneous fat into consideration. And the thickness of every single tissue must be also taken into consideration for the equivalent circuit calculation. Therefore, in our equivalent circuit for the epidermis, dermis and subcutaneous fat we added a capacitor in parallel. The values used for our calculations will be $\sigma = 0.2$ S/m and $\epsilon_r = 520\ k$, they are taken from [89].

Non-linear Electric Properties of Skin

The skin impedance has an electric current dependency and the non-linearity characteristics affects primary the conductivity of the skin layers by increasing the electric current for stimulation. Yamamoto and Yamamoto recognized that the non-linearity properties of the human skin originate in the skin, specifically in the keratin layer. We quote in their article, "*the skin itself can be considered as a cause for the non-linearity*" [92]. Keratin is a protein produced by keratinocytes, a predominant cell type in the epidermis found in the fifth layer of the epidermis, the stratum germinativum. Keratin is the key structural material that builds the outer layer of human skin. This protein migrates up from the third layer of the epidermis, the stratum granulosum, and fills the plate-like envelopes which the stratum corneum is composed [93].

The skin is formed by several layers (i.e. the epidermis, dermis and subcutaneous fat) in which, together, are the cause of the non-linearity. Therefore, in our equivalent circuit model, the nonlinear equivalent circuit will be placed between the equivalent circuit of the stratum corneum and the equivalent circuit of the lower layers (i.e. the rest of the epidermis, dermis and subcutaneous fat). In addition, when the electric current increases, the ionic conductance in this layer increases as well. Therefore, in our simulations, the nonlinear properties play an important role for the measurement of the electrode's voltage.

Equivalent Circuit of Biceps Brachii Muscle

The muscle that moves and supports the skeleton is called skeletal muscle. It consists of a number of muscle fibers lying parallel to one another and bundled together by connecting tissue. Muscle fibers are elongated and cylinder shaped measuring from 10 to 100 µm in diameter and up to 2.5 feet in length. The biceps brachii is characterized to be a flexor skeletal muscle and has high salinity and high content of water, which makes this muscle a good conductor. In addition, at low frequencies, the conductivity of muscle tissues dominates and at high frequencies, the relative permittivity tends to dominate [94]. Furthermore, the skeletal muscle presents anisotropic electric properties. It exhibits a higher longitudinal electric conductivity than the transverse conductivity. The reason is because the electric conduction is easier along the length of the muscle fibers than throughout the less conductive extracellular matrix. The next table illustrates the electric parameters used for the calculation. These parameters are measured at frequency of 10 Hz.

Table 3.2.1 Transversal and longitudinal electric properties of biceps brachii.

	Conductivity (S/m)	Relative Permittivity
Transversal	0.04 – 0.14 [89]	1.5 M – 40 M [89]
Longitudinal	0.2 – 0.5 [95]	10 M – 66 M [89]

For the model of the muscle is used just a resistor, because the resistive component dominates at low frequencies, and our simulations are within this range. For the calculations two different diameters are considered, 40 and 50 mm. Gomis [96] reports that the diameter of the bicep brachii in patients between 19 and 59 years old is between 29.4 and 51.2 mm.

Equivalent Circuit of the Cell Membrane

To model the activation of fiber cells the myelinated axon can be represented in a passive RC network, Figure 3.2.3. The membrane model is the passive membrane model used in the section 3.1. Here, R_i was incorporated, which is the intracellular resistance. $V_{e,n}$ is the extracellular potential at node n. $V_{i,n}$ is the intracellular potential at node n. C_m and R_m are membrane capacitance and resistance, respectively. V_r is the membrane resting potential, -70 mV from section 1.1.5

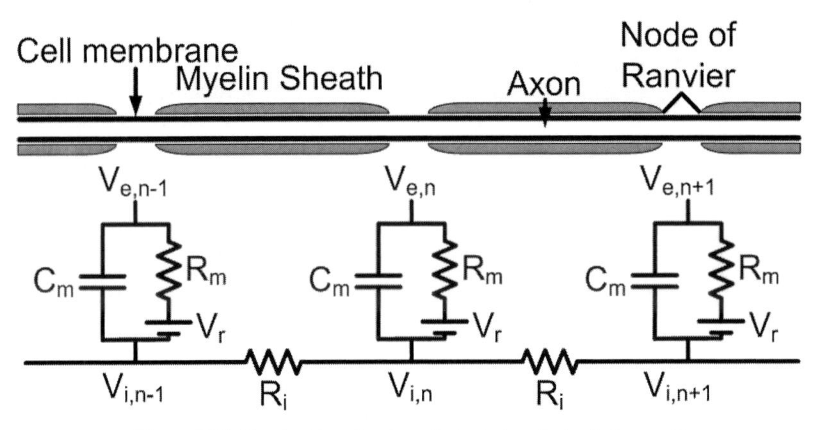

Figure 3.2.3 The electric network equivalent of a miyelinated nerve fiber.

The membrane current at node '*n*' is equal to the sum of the incoming axial currents and the sum of the capacitive and ionic currents through the membrane. The following equation describes the calculation of the transmembrane voltage [97].

Equation 3.7 $$R_i = C_m \frac{dV_n}{dt} + I_{i,n} = \frac{1}{R_i}\left(V_{i,n-1} - 2V_{i,n} + V_{i,n+1}\right)$$

The nerve becomes active when the second-order difference f (the activating function) of external node potentials V_e of a central node and its two neighbours exceeds a threshold (~ 20 mV), Equation 3.8 [35].

Equation 3.8 $$f = V_{e,n-1} - 2V_{e,n} + V_{e,n+1}$$

Meier [98] has shown this to be true for stimulation with a short rectangular current pulse of duration. There are several models for simulating the action potential propagation along the axon. However, many of them ignore the presence of voltage-gate channels in the membrane, and the potentials become smaller in amplitude and more spread out in time as they propagate away from the source. To deal this problem, there are approaches where the Hodgking-Huxley (section 1.1.9) model is used as membrane model by placing active elements to emulate the saltatory conduction [99].

The membrane parameters were calculated by considering a fiber with 20 μm of diameter. For the resistance were used a thickness of 3 nm [100], and a resistivity of 16 M $(\Omega * \mathrm{m})$ [101]. For the capacitance we will use 7 as relative permittivity of the membrane, it is typically between 2 and 10 [102]. For the calculation of the intracellular resistance these parameters were used, axoplasm resistivity 0.5 $(\Omega * \mathrm{m})$ [101] and a length of 1 mm was considered for the myelin sheath, from section 1.1.1.

Equivalent Circuit Model for NMES – Simulation

In the last sections, it was described each of the stages that integrates the model for NMES. The Figure 3.2.4 illustrates the equivalent circuit model which was simulated in PSpice. The equivalent circuit describes a 2D inhomogeneous tissue model.

The equivalent circuit of the stratum corneum was divided into two equivalent circuits. The reason was that the current density travels radially and not transversely; therefore, a resistance was placed between the equivalent circuits of the stratum corneum and between the electrodes to reproduce a more accurate simulation. This technique was used as well in the equivalent circuit of the epidermis, dermis and subcutaneous fat and biceps brachii muscle.

A resistance between the nodes of Ranvier is required. The longitudinally resistances in the muscle which are between the nodes of Ranvier were calculated with a length of 1 mm. The following tables illustrates the parameters used in the calculation of the passive elements (resistance and capacitance), for its simulation in PSpice.

Table 3.2.2 Parameters used in PSpice for the equivalent circuit model [80].

Layers	Transverse length (m)	Transverse area (m²)	Longitudinal length (m)	Longitudinal area (m²)	Transverse conductivity (S/m)	Longitudinal conductivity (S/m)	Relative permittivity
Electrode – Electrolyte	0.0005	0.0016	-		1/384	-	1
Stratum corneum	0.00002	0.0016	0.04	8e-7	1/90900	-	3.5 k
Lower layers	0.004	0.0016	0.04	160e-6	1/5	-	520 k
Biceps brachii musclee	0.05	0.0016	0.04	2e-3	1/7.14	1/5	8.3 M

Table 3.2.3 Calculated values for the cell membrane model [80].

Variable	Value
R_{mem} (Ω)	78.5 M
C_{mem} (F)	12.6 p
V_{rp} (V)	70 m
R_{axp} (Ω)	7.08 M
R_{rn} (Ω)	25

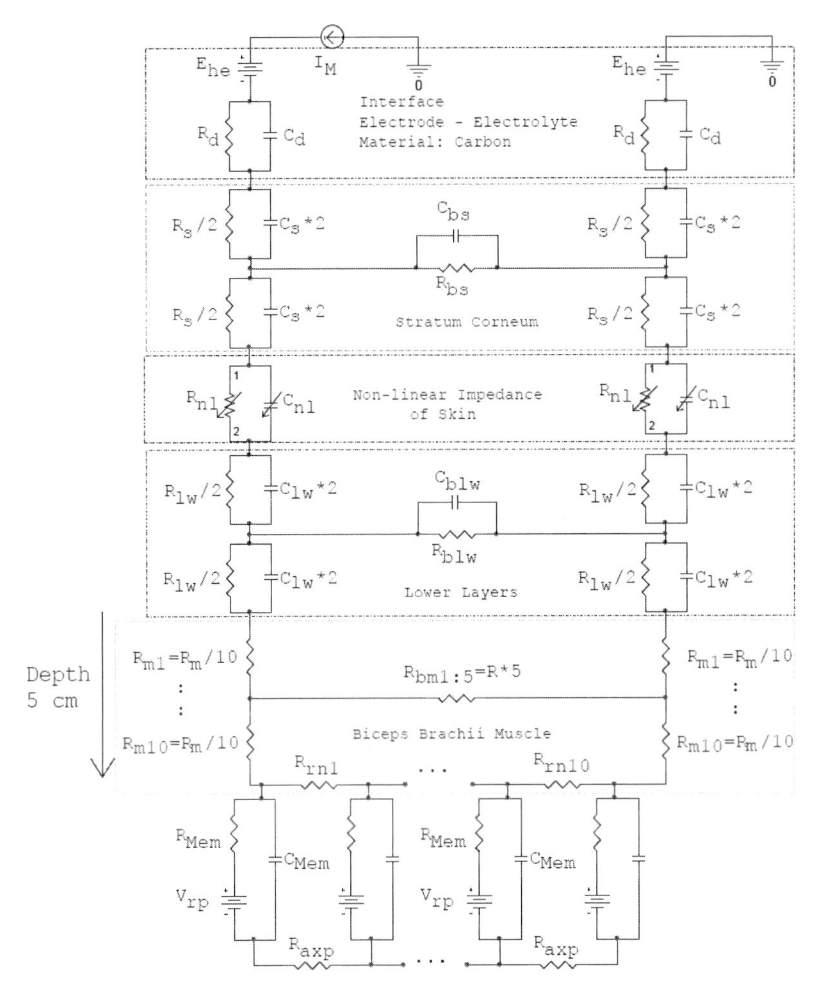

Figure 3.2.4 The Equivalent Circuit Model for the simulation of NMES. Lower layers represent the rest of the epidermis (without the stratum corneum), dermis and subcutaneous fat, reprinted from [80].

Incorporation of the Non-linearity Characteristics of Human´s Skin

By applying some current in the simulation, the voltage simulated needs to match the measured in the TENS experiment. The non-linearity problem was solved by adding variable elements in PSpice.

We gain knowledge that the conductivity of the skin increases when the electric current placed in the electrodes increases. The impedance of the model, as well, decreases by increasing the electric current. By changing the resistances in the model, the strength–duration curve almost stays constant; however, the electrode's voltage changes. To incorporate the nonlinearity characteristics in PSpice, resistances that deal with the non-linearity problem were placed in the equivalent circuit with the calculated values from the next equation [80]:

Equation 3.9 $$R_{nl} = \frac{|VTENS_{im} - VMODEL_{im}|}{i_m N_{Rnl}}$$

The variable $VTENS_{im}$ represents the voltage measured in the experiment in function of the i_m current applied, $VMODEL_{im}$ is the voltage in the simulation in function of the same current. The voltage in the simulation is measured before the incorporation of the non-linear elements. Finally, the N_{Rnl} variable represents the number of resistances placed in the model, in our case this value is 2.

In the Figure 3.2.4 the non-linear resistance is characterized by a variable resistor. In order to fulfill the value of this element in PSpice, we took the values measured in the TENS experiment, which range goes from 9.1 to 46.3 mA, and by using the last equation, we derivate the next one:

Equation 3.10 $$R_{nl} = 4178.28 * 0.9067399^{i_m}$$

For the incorporation of the non-linear capacitance, the charge of the entirely equivalent circuit model was the integration of the electric current in the model over the time, from 0 to t, where t is the pulse width; the voltage to-match ($VTENS_{im}$) in the TENS stimulation is needed as well, since the total non-linear capacitance should match the slope of the measured voltage in the TENS experiment. The reactance of the non-linear impedance is calculated by the equation $X = 1/(2\pi f C_{Tnl})$ where C_{Tnl} represents the total non-linear capacitance. The next equation was derived and the capacitances were calculated [80]:

Equation 3.11 $$C_{nl} = C_{Tnl} \times N_{Cnl} = \frac{\int_0^t i_m dt}{VTENS_{im}} N_{Cnl}$$

The variable N_{Cnl} represents the total number of capacitances placed in the equivalent circuit model, in our case 2.

3.2.3. Simulation with different Waveforms

In this section, different waveforms will be compared with against current, voltage, charge and energy requirements. The different waveforms used were: a monophasic rectangular pulse, a linear increase and sinusoidal.

The diameter of the biceps brachii was addressed in the section 3.2.2. The equivalent circuit of the biceps brachii muscle was calculated to include a depth of 50 mm. The diameters used in the simulation will be 40, and 50 mm. The reason for using two different diameters is because the musculocutaneous nerve is not in a straight line from the very start to the end; therefore, by using two different values we can analyze the physical depth of the musculocutaneous nerve.

In order to monitor the activation of the cells, the response of the cell membrane was measured with the transmembrane voltage as a function of the time, with the resting potential as initial value and increasing the voltage until the threshold was reached, -55 mV. The time when the transmembrane voltage overtakes the threshold was stored as pulse width. The following plot shows the strength–duration curves.

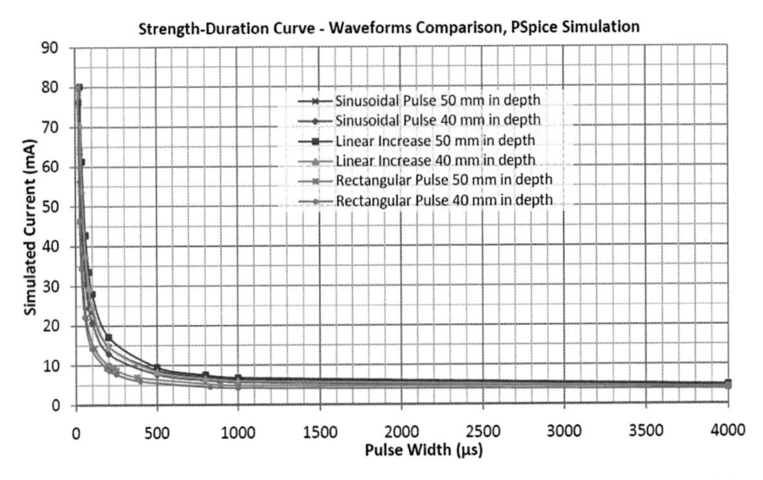

Figure 3.2.5 Comparison between strength duration curves between waveforms for simulation in PSpice, reprinted from [80].

On every point of the curve, the rectangular waveform requires less current amplitude than the others. However, there are more parameters in which it is important to pay attention. The necessary voltage at the electrodes is one of them. In our simulation, it is possible to observe that for pulse durations lower than 900 µs, the sinusoidal pulse at 40 mm in depth requires the lowest voltage. From 900 µs to 4000 µs, the linear increase pulse at 40 mm in depth requires the lowest voltage. For a depth of 50 mm; the sinusoidal pulse presents the lowest necessary voltage until 3000 µs. From 3000, the lowest necessary voltage is by stimulating with linear increase.

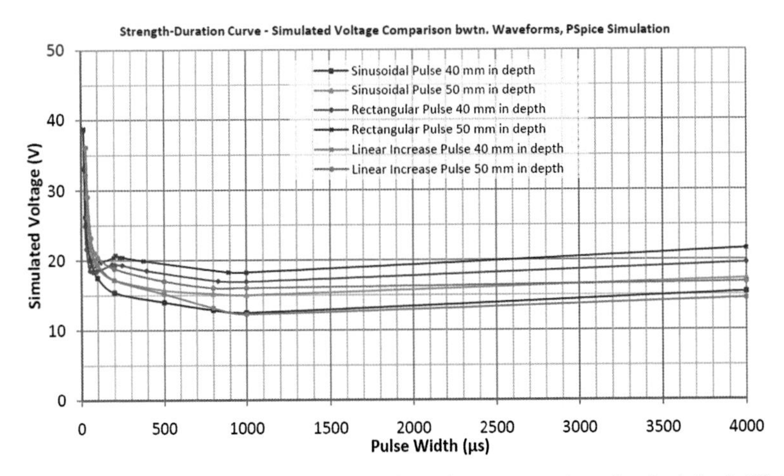

Figure 3.2.6 Comparison between simulated voltages between waveforms for simulation in PSpice, reprinted from [80].

Now, we will compare the electric energy and charge required to be delivered at electrode level. They were calculated only with the depth of 50 mm. The reason of calculating them at the deepest thickness was to measure the maximum electric energy and charge needed to stimulate the musculocutaneous nerve. The electric charge was calculated by integrating the amplitude of the current over the time. The following plot illustrates the threshold electric charge for the monophasic rectangular, linear increase and sinusoidal pulses versus the pulse width.

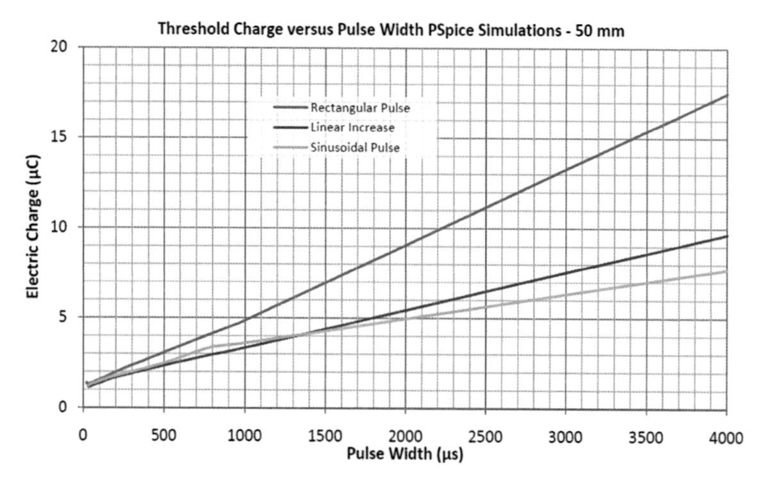

Figure 3.2.7 Charge required to reach the threshold of different waveforms versus the pulse width, reprinted from [80].

At short pulses, the linear increase pulse requires the lowest amount of charge, 1.14 μC at 30 μs; and for longer pulses, sinusoidal pulse has the lowest reaching 7.71 μC at 4000 μs. The rectangular pulse presents the lowest charge threshold when the pulses are evaluated at their own chronaxie time.

The electric energy was calculated by integrating the multiplication of voltage and current over time. The required energy at electrodes to reach the membrane threshold is plotted in the Figure 3.2.8. Linear increase has better results than monophasic rectangular and sinusoidal pulses, with the minimum electric energy of 17.5 μJ at 80 μs. When the three pulses are evaluated at their chronaxie value, linear increase has the lowest energy reaching 23 μJ at 475 μs. Sinusoidal pulse has the lowest energy at long durations from 1900 to 4000 μs.

The next table shows the results when the simulated waveforms are evaluated at their chonaxie values. The minimum values for each evaluation are in bold.

Table 3.2.4 Charge and Energy of three different waveforms evaluated at their chronaxie [80].

Simulated Waveform	Chronaxie (μs)	Charge (μC)	Energy (μJ)
Rectangular pulse	**248**	**2.17**	33
Linear increase	475	2.3	**23**
Sinusoidal	350	2.35	26

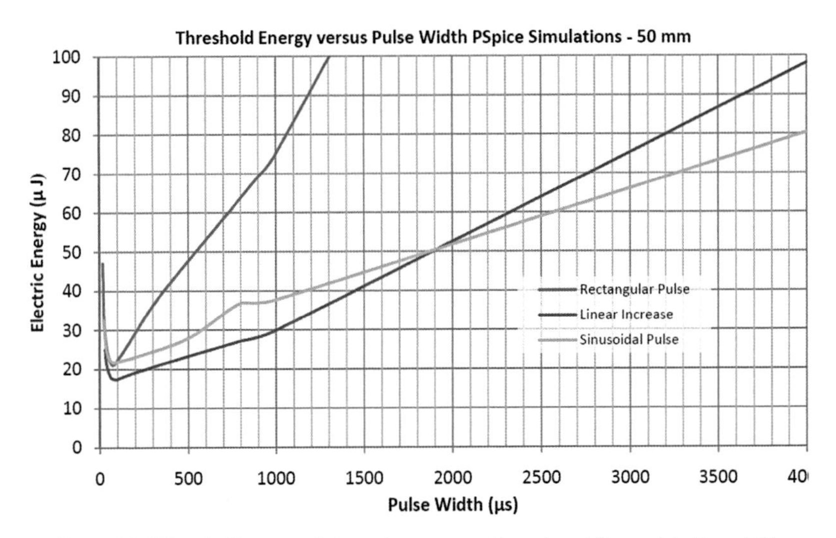

Figure 3.2.8 Threshold energy of al waveforms versus the pulse width, reprinted from [80].

3.2.4. Discussion

The model was tested under different stimulation waveforms. By summarizing the simulations, the rectangular pulse has better results in electric charge when it is evaluated at its chronaxie value. Regarding the electric energy, the linear increase shows better results when is evaluated at its chronaxie value. In our case, comparing the monophasic rectangular, sinusoidal and linear increase pulse, the latter stimulus waveform seems to be the more appropriate pulse since it minimizes the electric energy for the simulation at the chronaxie value.

This model serves to calculate the parameters needed for stimulating the musculocutaneous nerve. This insight can be used for manufacturing effective devices in which the energy or voltage amplitude is a limitation.

The outcomes of the minimum electric energy and charge gave us the information on which waveform could be appropriated to stimulate the musculocutaneous nerve. However, an *In Vivo* experiment by stimulating with different waveforms is shown in the section 3.4, it was performed in the order to corroborate the obtained results and further analysis is described.

3.3. Portable Stimulator

In this section is described the portable neuromuscular stimulator that we published on [103], which was developed for performing In Vivo experiments with different waveforms. The system is programmable and can deliver current signals, it can be completely adjusted to perform several experiments or it can have fixed parameters to be used in rehabilitation and in daily use for FNS.

The device is battery powered and can be controlled wirelessly. The adjustable current output is one of the advantages of this stimulator, because it can deliver several waveforms and it is possible to change every parameter on the stimulation signal.

Stimulation current signals formed by pulse trains can be defined, parameters like waveform, amplitude, pulse width, interphase interval between positive and negative phase or time between pulses can be configured by the user.

The system can be divided into three components: the first one is a portable stimulator, which performs the electrical stimulation; the second one is the USB base station (hereinafter referred to as base or base station), which functions as an intermediate and communicates the PC and the portable stimulator; the third one is graphical user interface, which enables the user to control and monitor the stimulation.

3.3.1. The stimulator

The device is powered by a battery. On one hand the battery is important because of the safety reasons, so the stimulating device, which comes in contact with the human body, is not connected to the power grid, on the other hand because of the mobility and the flexibility of the portable stimulator.

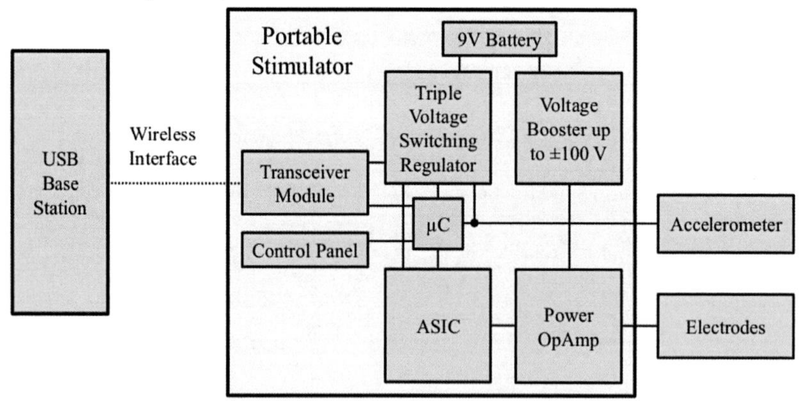

Figure 3.3.1 Block diagram of the portable stimulator.

A microcontroller (PIC18F46J50, Microchip) was used as a processing unit and programmed for the portable stimulator.

The core of the stimulator is the ASIC version 1 presented in the chapter 4. This ASIC behaves like a 9 bit DAC. It generates an analog signal from a digital data which is received in serial way and stored in an internal memory. The ASIC has one bipolar current output channel. An amplifier is connected directly at the output of the ASIC in order to deliver a current signal with amplitudes up to ±100 mA.

Due to the low voltage supply and the necessity of higher voltages to perform transcutaneous stimulation a voltage booster was implemented, it receives 9 V at its input and delivers ±100 V at its output.

Several electrodes can be used to conduct electrical current to the skin. Two universal 4 mm bunch pin plugs terminate the cable line of the portable stimulator in order to connect the electrodes.

In case of experimentation an external three-axial accelerometer (LIS344AL, ST) is attached to the stimulator. By locating it on any joint it enables the measurement of the muscle reaction or even the angle change of the joint because of the stimulation.

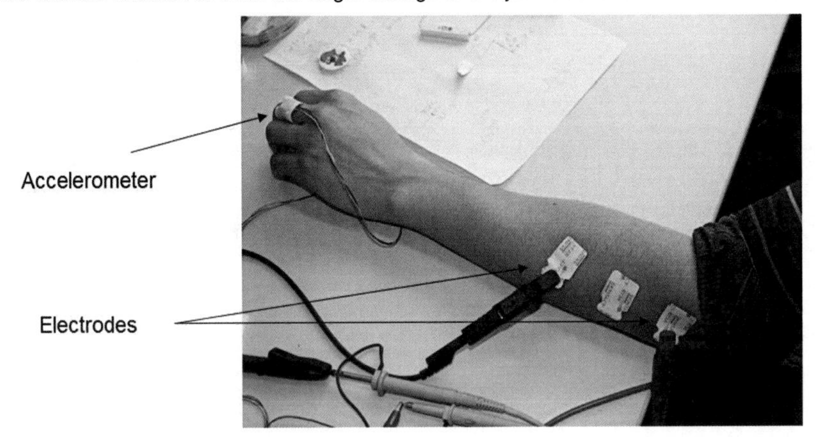

Figure 3.3.2 Example of the accelerometer and electrodes collocation.

The control is effected either by the PC via the wireless interface or by the control panel. The control panel is composed by four push-buttons. The buttons can be programmed individually to perform individual tasks, i.e. to change the stimulation waveform, change the intensity of the output current, change the pulse width, etc.

A wireless transceiver module (MRF24J40MA, Microchip) based on the IEEE Std. 802.15.4 is implemented to establish the link with the PC through the base station.

The portable stimulator is enclosed in a plastic box (16x8x2.5cm³) and is powered by a 9 V rechargeable battery. The weight of the portable stimulator is 272.60 g with the battery.

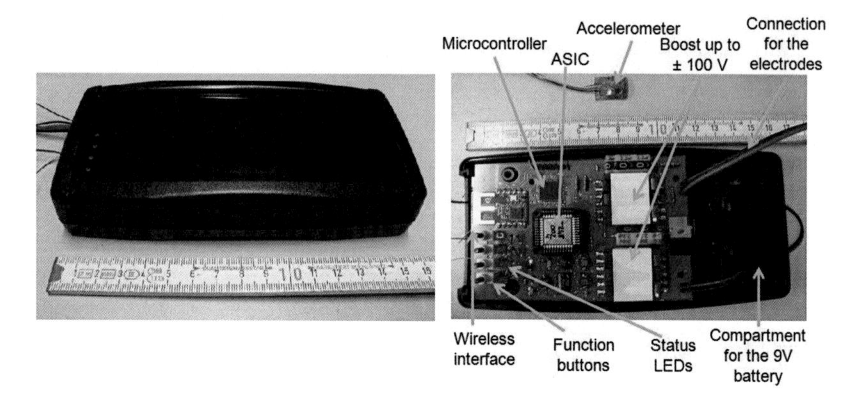

Figure 3.3.3 Picture of the portable stimulator.

The device can deliver a signal with a maximum amplitude of ±100 mA, but by changing the gain resistor of the amplifier the output was limited to ±66 mA because of safety reasons. The output voltage will depend on the electrodes impedance, the device is able to handle up to ±100 V.

By stimulating with a continuous signal the current consumption of the system was found to be around 130 mA. For the measurement the stimulator was programmed to deliver continuously a sinusoidal signal. The parameters were: 33.5 mA amplitude, 290 µs pulse width, without interphase and 22 ms interpulse. At its output the two electrodes were connected at the forearm of a person.

According to the manufacturer the wireless control has a range of 400 ft (121.92 m). We tested the connection indoors for short distance communication without problems.

The size and weight of the stimulator make it suitable for daily use as FES. However, its autonomy is too short for the application, e.g. according the measurements it sinks around 130 mA. If we consider the use of a rechargeable battery which is commercially available, 9.6 V with a capacity of 200 mAh, it would function continuously around 1.5 hours without a change of battery. However, as shown its flexibility makes the device suitable for perform experimentation with changing parameters.

3.3.2. USB base station

The base station (PIC18 Explorer Board, Microchip) is the bridge between the PC and the stimulator. It is connected to the PC through its USB port. A daughter card (MRF24J40MA PICtail, Microchip) is attached in order to establish the wireless link with the stimulator. Its core (PIC18F87J11, Microchip) was programmed to perform the corresponding tasks.

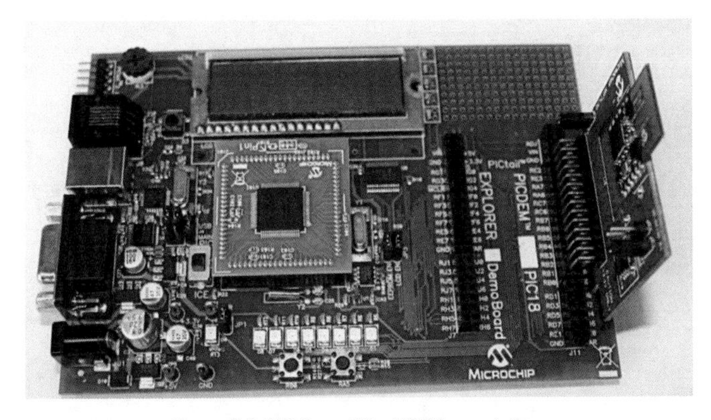

Figure 3.3.4 Picture of the USB base station.

3.3.3. Graphical User Interface

Different graphical user interfaces (GUI) were developed to perform different tasks.

The first version program was developed for manual control over the portable stimulator. Visual Basic was used as application development tool. The user can manually generate the current signals, it is possible to choose the waveform, amplitude, interphase interval and time between pulses. It is also possible to perform either a single or a continuous stimulation, as well as to change the amplitude during the experiment. The interface makes also possible to see the response of the accelerometer.

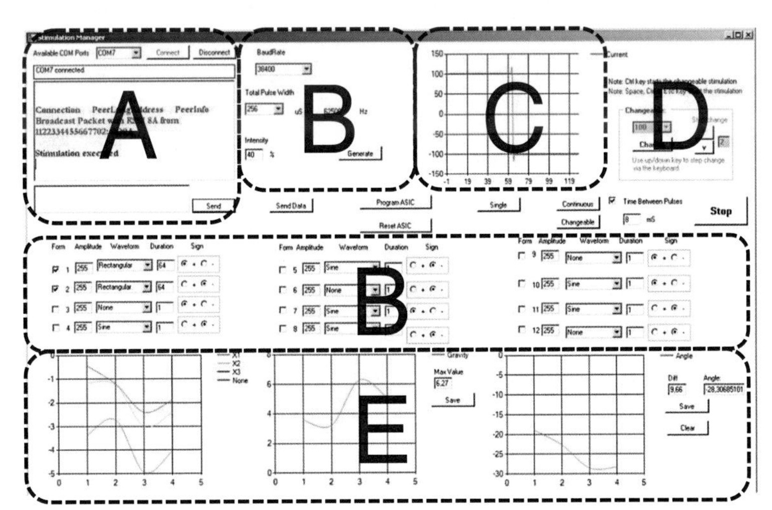

Figure 3.3.5 Visual Basic front panel of the GUI, a)status windows, b)stimulation parameters, c)generated stimulation waveform, d)command buttons, e)monitor of the accelerometer.

An interface in Matlab was also implemented, with this is possible to control the stimulator directly from algorithms running in online in Matlab. The portable neuromuscular electrical stimulator can be combined with EMG/EEG biosignal acquisition methods, in order to perform the bypass in patients which the neural pathway is broken or interrupted by any disease or accident.

Another version of the user interface was developed with LabVIEW especially to run automatically experiments, section 3.4, by changing one or more variable, see Figure 3.3.6. This GUI enables precise experimentation due to its ability of loading a preconfigured file. The file should contain the number of signals to deliver on every experiment. All the parameter can be adjusted individually for every signal, e.g. it makes possible to perform an experiment by sweeping only a parameter or a combination of all of them. The signals can be single pulse or continuous. Figure 3.3.7 shows the adjustable parameters.

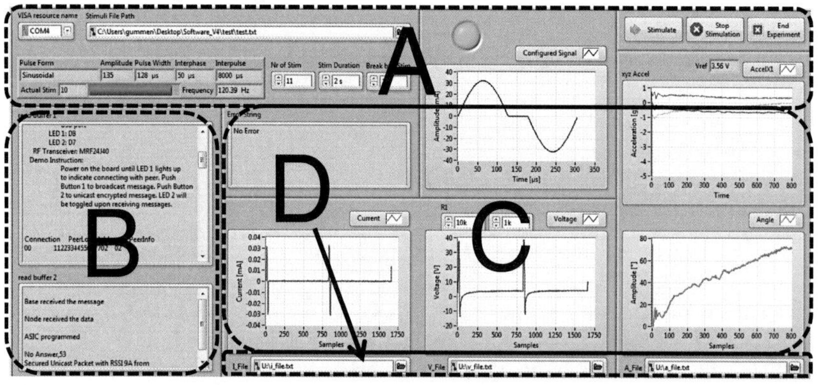

Figure 3.3.6 Labview front panel of the GUI, a) configured parameter (pulse form, anodic/catodic, phase width, interphase, interpulse interval, stimulation duration, number of stimulation pulses, time between individual stimulation) and experiment control buttons, b) status windows, c) monitor of the acquired data, d) path file to save the data.

The interface also controls a data acquisition card (DAQ M PCI-6289, National Instrument) on the PC. The card is connected to the electrodes to perform the voltage and current measurements.

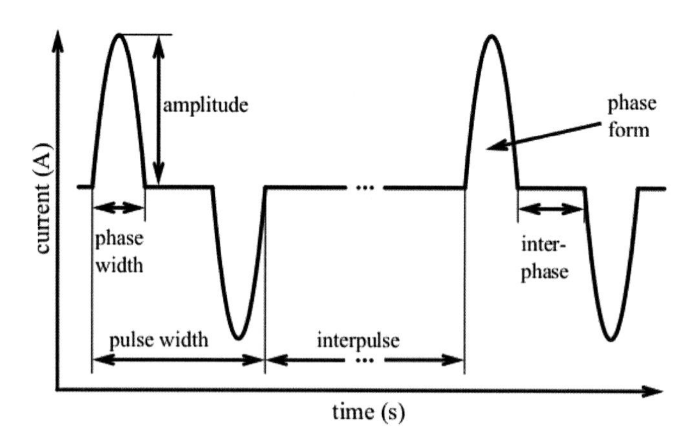

Figure 3.3.7 Adjustable parameters of the output signal.

3.4. In Vivo Experimentation

In this section we analyze the properties of using different waveforms for NMES, we published the results in [104]. The experiments were done by applying a single pulse (rectangular, sinusoidal, linear increase or linear decrease) and by measuring the muscle reaction with an accelerometer. For practical applications it would be required to perform stimulation with a pulse train, but the scope of this study is to compare the properties of different waveforms. The reasons for using a single pulse is to trigger only a single action potential and to leave out others parameters as interphase, distance between pulses or width of the train. The use of an accelerometer gives a quantitative metric to compare the muscle reaction against stimuli with the different waveforms.

3.4.1. Material and Methods

Subjects for In Vivo Experiments

The In Vivo experiments were performed under written consent of 2 healthy subjects. In all the cases there was no knowledge of neurological or orthopedic disease history, and it is assumed that the involved muscle is innervated.

The experiments were conducted over five days with the same test subjects and the data were averaged.

Experimental Procedure

For the stimulation we used our portable stimulator described in section 3.3 and it was published in [103]. For safety reasons the device is battery powered, and it is controlled through the PC via wireless IEEE Std. 802.15.4 interface. It can deliver analog current stimulation pulses from a digital data stream, the maximum output voltage is ±100 V and the current is limited to ± 66 mA.

To contact the skin we used two fully gelled electrodes 45 x 80 mm² (PG473, FIAB). The impedance of the electrodes was measured before and after the experiments with an LCR meter (4284A, Agilent Technologies).

A 1 Ω resistor was connected in series to measure the current flowing through the electrodes. The electrode voltage was measured by connecting in parallel a resistive divider formed by a 10 kΩ and a 1 kΩ resistor. In order to measure the muscle twitch response a three-axial accelerometer was used (LIS344AL, ST). The data was acquired by the computer via a National Instrument DAQ M 6289 PCI card at sampling rate of 100 ksps per channel.

The subjects were seated comfortably in a chair, and after cleaning the skin surface, the electrodes were placed on the right forearm. The reference electrode was positioned on the elbow over the ulnar nerve and the working electrode over the extensor muscle, with 10.50 cm of separation. The extensor muscle was found by a visual inspection during the movement of the middle finger. The accelerometer was placed on the proximal phalange of the middle finger. The forearm was fixed on the armrest of the chair with the hand hanging in order to avoid the contact with any surface, Figure 3.4.1. The subjects had a book to read and keep them distracted from the experiment.

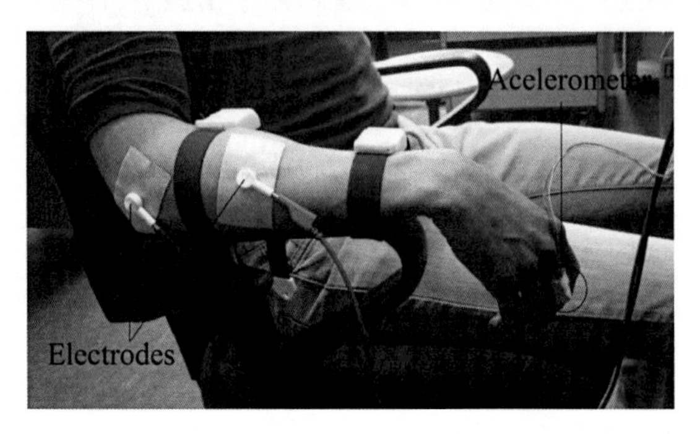

Figure 3.4.1 Position of the arm during the experiment.

We used Labview for the experiment, the software application takes a generated stimuli file with the waveforms, pulse widths and amplitudes of the stimulation signals, and starting from the first pulse, send the information through the wireless interface to the stimulator, triggers the stimulation and simultaneously through the Nationals Instrument card acquires the current, voltage and inertial forms.

The pulses were symmetrical biphasic current pulses (anodic first) without interphase, Figure 3.4.2. The used waveforms were: rectangular, sinusoidal, linear increase and linear decrease. The pulse widths were: 128 µs, 256 µs, 512 µs and 1,024 µs. The amplitudes were swept from 7.7 to 66 mA with steps of 6.5 mA. The number of signals for a complete sweep of these values was 160, each one was repeated 5 times in order to take the average of the results, then the total of pulses were 800, with a separation of 1.5 s, the experiment with each subject lasted 20 minutes. The stimuli were randomly organized in order to avoid inhibitory effects of the subject because of the expected pulse.

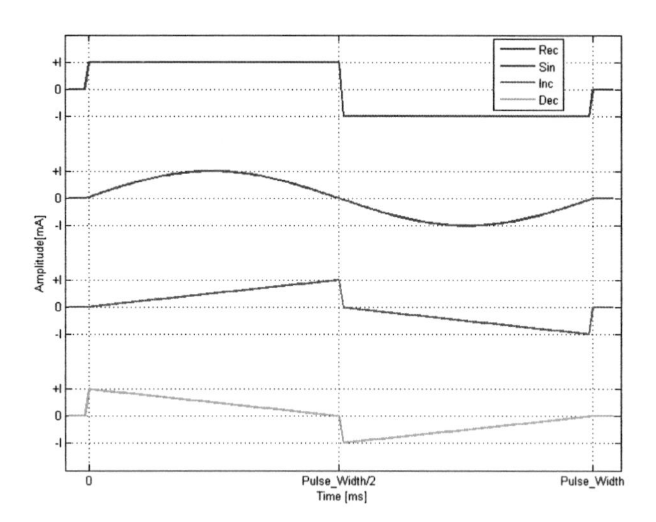

Figure 3.4.2 Waveforms used for the experiment.

Data Analysis

In the Labview environment the data for the voltage and current was limited to the pulse width, and the data of the three axis of the accelerometer was firstly filtered with a Butterworth 3th order lowpass filter, then down-sampled to 400 sps and limited to 500 ms.The three axial accelerometer gives the acceleration in a specific direction. Thus, the data was converted to spherical coordinates from which was taken only the radius, i.e. the amplitude without direction.

A Matlab script organized the data and performed some tasks in order to obtain the desired curves, such as the current and voltage peaks in all the samples. The charge delivered by the stimulator in each case was obtained by integrating the current over time. The energy delivered was calculated by integrating the multiplication of voltage and current over time. The inertial response was used as measurement of the muscle reaction time, only the first peak of the acceleration data was taken into account, because it corresponded to the muscle twitch and the second and in some cases the third, corresponded to the movement of the accelerometer to the initial position. For each configuration parameter, there were five samples taken and the lowest and highest values were neglected, and the three other values were averaged. Mathematical interpolation was performed on the data for obtaining the strength duration curve at a given inertial response. When the accelerometer is at rest its output is 1 gravity (g). Thus, in order to measure weak and strong muscle reaction, the given inertial responses for the experiment were 1.2 g and 1.8 g.

The data obtained were recorded on a daily base. Using the values recorded over five days, the average and standard deviation were calculated for every subject.

3.4.2. Results

Figure 3.4.3and Figure 3.4.4show the results for the measurements corresponding to the subject 1 and subject 2, respectively.

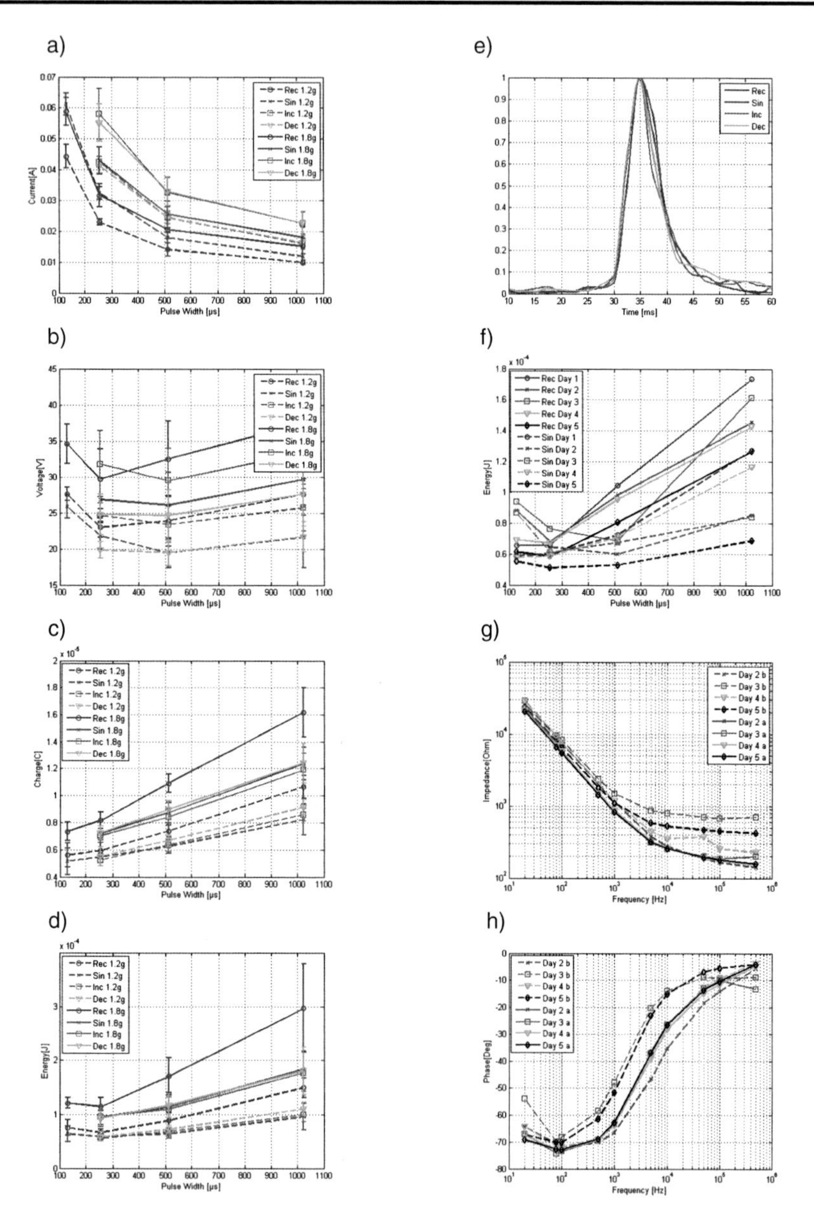

Figure 3.4.3 Measurements of subject 1 with different waveforms, a) strength duration curve, b) required voltage peak, c) required charge injection, d) required energy at the electrodes, those four graphs show the necessary values in order to reach 1.2 g and 1.8 g, e) distribution of muscle reaction time, f) required energy in order to reach 1.2 g with rectangular and sinusoidal waveforms on different days, the last two curves are the impedance at electrodes on different days before and after the experiment, g) magnitude and h) phase.

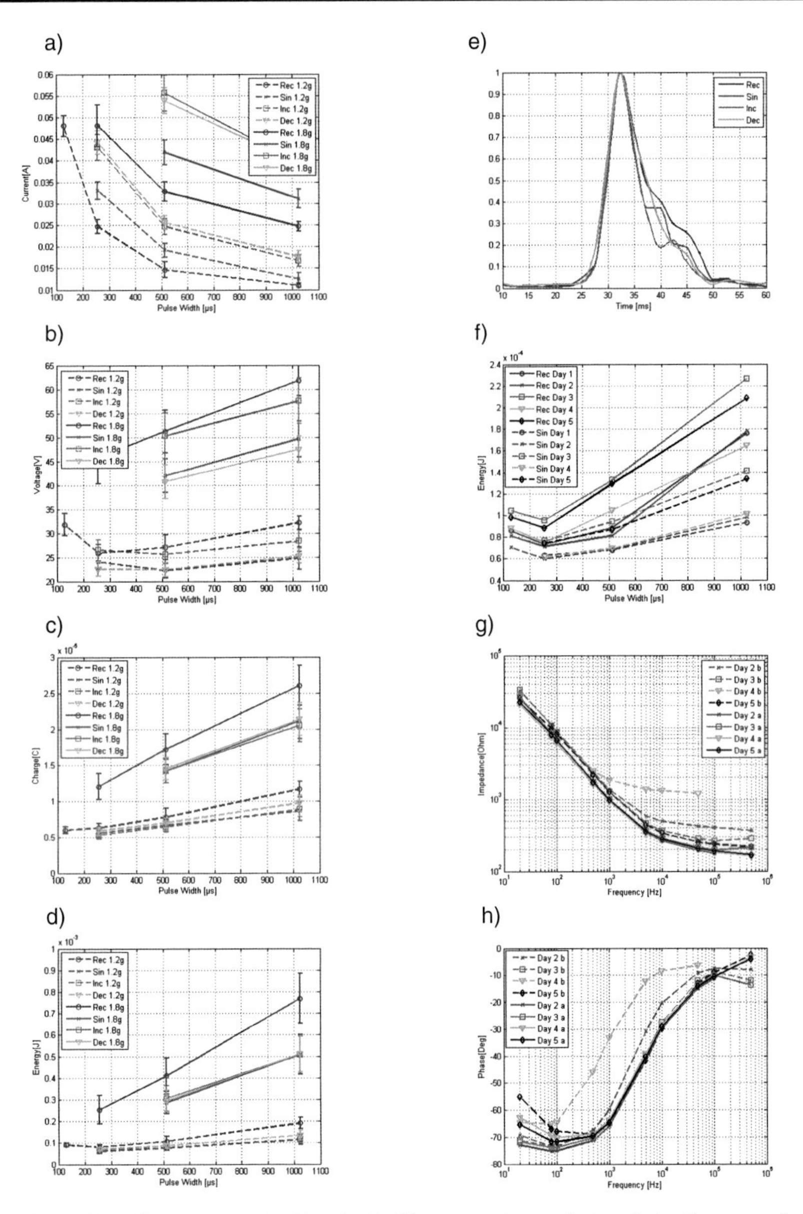

Figure 3.4.4 Measurements of subject 2 with different waveforms, a) strength duration curve, b) required voltage peak, c) required charge injection, d) required energy at the electrodes, those four graphs show the necessary values in order to reach 1.2 g and 1.8 g, e) distribution of muscle reaction time, f) required energy in order to reach 1.2 g with rectangular and sinusoidal waveforms on different days, the last two curves are the impedance at electrodes on different days before and after the experiment, g) magnitude and h) phase.

The first four subfigures show the results for an inertial reaction of 1.2 g and 1.8 g against stimuli produced by different waveforms. Subfigure a) shows the strength duration curve, i.e. the necessary current peak in order to reach the desired reaction. Subfigure b) shows the required voltage peak. Subfigure c) corresponds to the required charge injection at electrodes. In subfigure d) are the energy levels required at the electrodes. Subfigure e) shows the distribution of the muscle reaction time for stimulation with the different waveforms. Subfigure f) shows the required energy in order to reach 1.2 g with rectangular and sinusoidal waveforms on the different days. The last two subfigures correspond to the electrode impedance on the different days before and after the experiment; subfigure g) magnitude and h) phase.

Because of the current limitation on the stimulator, in some cases was not possible to reach the required current level, i.e. for subject 1 with a 128 µs pulse width; and for subject 2 with 128 µs and 256 µs pulse widths.

The experiments with subject 2 showed the same tendencies with similar values for the 1.2 g reaction, but different values for the 1.8 reaction, i.e. the threshold values were higher.

Figure 3.4.5 shows current and voltage measurements for stimulation with a 256 µs pulse width on subject 1.

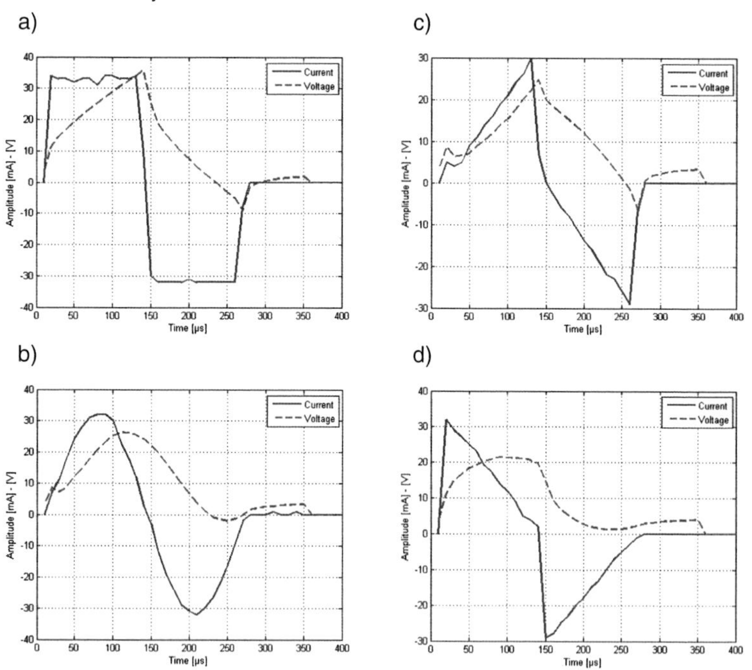

Figure 3.4.5 Current and voltage waveforms measurements with a 256 µs pulse width on subject 1 for different waveforms; a) rectangular, b) sinusoidal, c) linear increase and d) linear decrease.

Figure 3.4.6 shows the answer of the accelerometer for the stimulation of Figure 3.4.5 a).

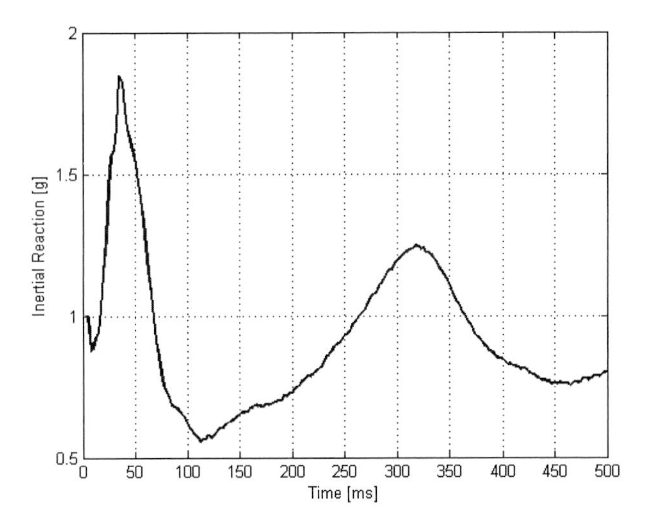

Figure 3.4.6 Response of the accelerometer for stimulation with a 256 µs pulse width.

The Table 3.4.1 shows the required current peak to achieve two different magnitudes of muscle reaction and the respective percentage for the different waveforms by stimulating with a pulse width of 256 µs. It can be seen that the tendency is the same for the two subjects.

Table 3.4.1 Required current with a 256 µS pulse width.

	Rect[mA]	***Sin[%]***	***Inc[%]***	***Dec[%]***
Subject 1, 1.2g	23.01	140.93	185.24	179.52
Subject 2, 1.2g	24.68	133.85	174.44	179.19
Subject 1, 1.8g	31.71	135.50	182.91	174.23
Subject 2, 1.8g	48.11	NA	NA	NA

* The values of the different waveforms are related to the rectangular pulse corresponding to the same subject and same reaction.

The Table 3.4.2 shows the required voltage peak to achieve two different magnitudes of muscle reaction and the respective percentage for the different waveforms by stimulating with a pulse width of 256 µs. The tendency is the same for the two subjects.

Table 3.4.2 Required voltage with a 256 µS pulse width.

	Rect[V]	***Sin[%]***	***Inc[%]***	***Dec[%]***
Subject 1, 1.2g	23.00	94.97	107.09	86.30
Subject 2, 1.2g	25.92	92.90	102.06	87.03
Subject 1, 1.8g	29.71	90.37	107.11	83.42
Subject 2, 1.8g	45.76	NA	NA	NA

* The values of the different waveforms are related to the rectangular pulse corresponding to the same subject and same reaction.

The Table 3.4.3 shows the required energy to achieve two different magnitudes of muscle reaction and the respective percentage for the different waveforms by stimulating with a pulse width of 256 µs. The tendency is the same for the two subjects and in all the cases there is reduced energy consumption.

Table 3.4.3 Required energy with a 256 µS pulse width.

	Rect[µJ]	*Sin[%]*	*Inc[%]*	*Dec[%]*
Subject 1, 1.2g	67.18	88.11	84.33	88.64
Subject 2, 1.2g	80.66	83.13	77.19	88.54
Subject 1, 1.8g	114.72	82.84	84.41	81.13
Subject 2, 1.8g	254.14	NA	NA	NA

* The values of the different waveforms are related to the rectangular pulse corresponding to the same subject and same reaction.

3.4.3. Discussion

The values 34.88 ms and 32.38 ms, correspond to the center of the distribution curve of muscle reaction time for subject 1 and subject 2, respectively. Despite there is a difference between subjects, there is not a significant difference for the waveforms on the same subject, i.e. the stimulation with different waveforms does not affect the muscle reaction time.

Despite the fact that the threshold levels are not the same for both of the subjects, the tendency regarding the necessary peak current remains. The waveform for the lowest peak current is rectangular followed by sinusoidal, linear decrease, and linear increase. However, the two last waveforms require almost the same current because of their nature, the waveform is the same but inverted. The tendencies are the same for the 1.2 g reaction and for the 1.8 g. Nevertheless, in the latter case the distance between rectangular and the different waveforms is higher for the subject 2, e.g. by stimulating with a 1024 µs pulse width, the sinusoidal signal require 13% more current than the rectangular for the 1.2 g reaction and 25% for the 1.8 g curve, respectively.

The voltage curves show also the same tendency for both subjects. As expected, in most of the cases the voltage peak necessary to reach the same muscle reaction is lower for different waveforms than for rectangular.

The energy required in all the cases is lowered by stimulating with waveforms other than rectangular. The optimal point would be located at different points, depending on the desired waveform and how strong the expected reaction is. The optimal pulse width is found to be around 256 µs, where the chronaxie is. In the case of 128 µs pulse width the required current is between 85% and 95% higher, the voltage between 16% and 22% higher, the charge between 4% and 10% lower and the energy between 5% and 13% higher. In case of 512µs the required current it would be between 30% and 45% lower, the voltage between 11% lower and 12% higher, the charge between 14% and 42% higher, and the energy between 9% and 61% higher.

By stimulating with a 256 µs pulse width, each waveform shows its own attributes. The waveform for minimum current is the sinusoidal, but the curve with the lowest voltage requirement is linear decrease, and the waveform which requires the least amount of energy is linear increase. However, between these four waveforms, sinusoidal and linear decrease seem to be the more balanced options, both of them require lower voltages than rectangular, the energy required is also lower than rectangular but similar between them, there is only a trade-off depending the requirements of the system, lower current or lower voltage.

In the curves, it can be seen that the standard deviation of the required energy for rectangular pulse sometimes overlaps the standard deviation of the others waveforms. However, it not necessarily means that the required energy for the different waveforms is sometimes lower than the required for rectangular waveform and sometimes not. By analyzing the Figure 3.4.3 and Figure 3.4.4, in subfigures f) it can be said that the use of different waveforms was more efficient than the rectangular form, but the threshold level was different every day. Thus, the subfigures g) and h) show the magnitude and phase of the impedance for the different days. Before the experiment the measurements were different every day but the impedance was already stable after the experiment and the values are similar regardless of the day. Despite the impedance before the experiment was higher on the days associated to lowest energy, there is no direct relation and no logic explanation. The issue of the variation of the stimulation threshold levels is related to the electrodes position displacement on the days.

The charge injected to the electrodes is a metric which permits the calculation of the energy required for the whole system. It is necessary to know the architecture of the output stage. In stimulator systems with current steering architecture, the total energy could be obtained, by multiplying the required charge at the electrodes by the voltage of the system.

3.5. Chapter Conclusion

With the use of a passive membrane model, it was shown that energy can be saved by stimulating with non-rectangular waveforms. However, in order to have more accurate results, a second model was developed to incorporate equivalent circuits for the electrodes and tissues along the path.

The second model is an equivalent circuit of the elements in the path from the electrode to the cell membrane. This equivalent circuit model was divided by tissues, the aim of which is to facilitate future research analysis of biological models. The removal of sub-equivalent circuit models can be done and can simulate nerve stimulation from another specific point, e.g. below the stratum corneum.

A setup for experiments was developed, i.e. portable neuromuscular stimulator and its graphical user interface. Its flexibility makes the device suitable for performing experiments with changing parameters. The system enables the experimentation not only with the four mentioned waveforms, but also allows experimentation with continuous stimulation pulses and the comparison of different waveforms against the changing of parameters such as interphase, distance between pulses or width of the train. The size and weight of the stimulator make it suitable for daily use as FES. And its design serves also as basis for the further development of a multielectrode stimulator.

By *In Vivo* experimentation, the results of single pulse stimulation show more efficient neuromuscular stimulation by using different waveforms rather than the rectangular pulse, which is nowadays the most widely used waveform. The tendency of the curves is consistent over several days, using multiple test subjects. The required energy is lower in all cases. The voltage is also lower in most cases. Each waveform exhibits its own attributes. The waveform could be selected according to the specifications of the system and according to the requirements for either lower currents or lower voltages.

The optimal pulse width was found to be around 256 μs, according the current, voltage, charge injection and energy requirements.

The required charge injection at the electrodes was lower by using non-rectangular waveforms. By knowing the architecture of the stimulator output stage, the charge injection requirement is a metric that permits the calculation of the energy required by the whole system.

4. Circuitry of a Neurostimulator

In this chapter, we show our design of three ASICs, each one is an enhanced version of the previous one, we published the results in [105] [106] [107] [108]. They are neurostimulators, intended to drive microelectrodes small enough to perform invasive electrical stimulation.

The system shall fulfill certain characteristics according to the application. In some cases it is desired to have fully implantable devices, making possible its use in daily activities while providing comfort to the patients. For fully implantable devices, size plays an important role. Thus, we have chosen a 130 nm CMOS process as an economical compromise between minimum feature size and cost. Furthermore, implantable devices are usually powered by an inductive coupling and the energy is limited. Therefore, it is necessary to implement low power circuit designs and to combine different methods to save energy during the stimulation.

It is desirable to stimulate small groups of neurons or even single neurons. Thus, it is convenient to build small electrodes in order to achieve high selectivity. However, with smaller electrodes, higher stimulation voltages are required, due to the electrode impedance. Through current stimulation with different waveforms it is possible to reduce the voltage peak of the injected signal necessary to achieve the firing of an action potential.

In the chapter 3, it was seen that it is possible to obtain the same stimulation reaction with lower voltage peaks and lower energy thresholds by stimulating with non-rectangular waveforms, because of it, the designs presented in this chapter are capable to stimulate with several waveforms in order to save stimulation energy.

It is also important to maintain low power dissipation, since the increase of temperature in tissues or brain could be harmful.

The stimulation signals could be either current or voltage signals, but most commonly used are current signals since the natural stimulation is performed through electrical current. Besides, current pulses are preferred over voltage pulses to eliminate variations in the stimulation threshold as a result of the changes in the electrode-tissue impedance.

4.1. ASIC Version 1

In this section, a design that we published in [105] [106] is shown. The developed ASIC is composed of a stimulator, capable of driving several current waveforms, and an analog channel for biosignal acquisition. The ASIC enables experimentation by stimulating neurons with different current waveforms, amplitudes and frequencies. There is also an integrated analog channel for biosignal acquisition in order to analyze the response of the neurons against different stimuli, without necessity of extra devices.

4.1.1. ASIC Description

The stimulator consists of a serial input interface, an 8 bit DAC, and an output stage capable of driving bipolar current signals. The analog channel is composed of an AC-coupled preamplifier, a lowpass filter and a postamplifier.

Stimulator

The structure of the circuit is shown in Figure 4.1.1. The information is first received in digital form through the serial port *data_in*. By using a serial interface it is possible to save silicon area, because IO pads require relatively large areas of silicon.

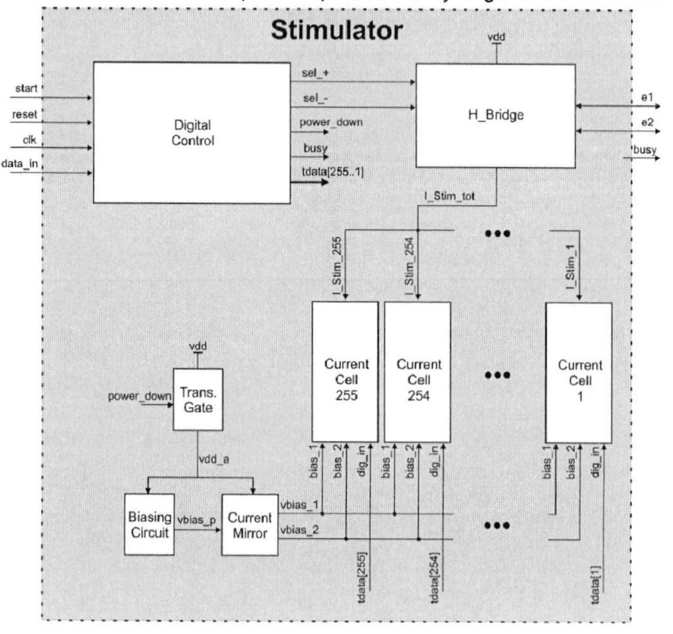

Figure 4.1.1 Block diagram of the stimulator, ASIC v1.

The "*Digital Control*" module has internal memory to store the setup configuration, once the digital module is programmed, it is used as a controller of the analog stage. The module has a data bus output to send information to the DAC which is composed of 255 "*Current Cells*" which are responsible for converting the digital data into an analog current signal. The 8 bit DAC is able to source up to 256 different current values; it also has a *power_down* line to put the "*Voltage Bias*" circuit in standby mode in order to save power; there is also an "*H_Bridge*" circuit which enables the bipolar output by inverting the polarization of the output lines, and makes it possible to isolate the output pins of the output stage for performing other tasks, such as biosignal acquisition. The stimulator behaves like a 9 bits DAC, because it allows up to 256 levels of positive current and 256 levels of negative current.

Stimulator, Digital Control

The block diagram of this module is illustrated in Figure 4.1.2. It receives the DAC sample codes serially, stores them in parallel and drives them out in thermometer code. All the digital logic reacts to rising edge clock, and its inputs and outputs are active high, except *power_down*.

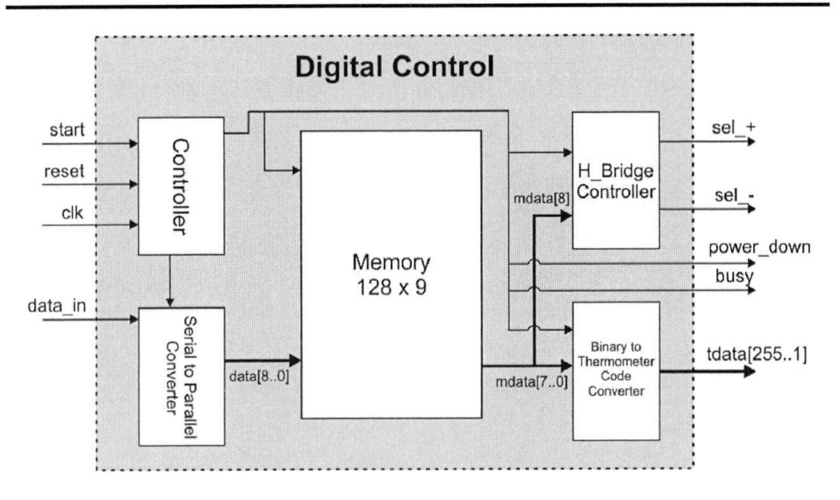

Figure 4.1.2 Block diagram of the digital control, ASIC v1.

When this module is not busy, it is put into programming mode by receiving a bit string with the value "*101*" on its *data_in* line; after that, it receives 128 words of 9 bits each in consecutive order, Less Significant Bit (LSB) first as shown in Figure 4.1.3. The format of the words should be handed in Sign-Magnitude Representation.

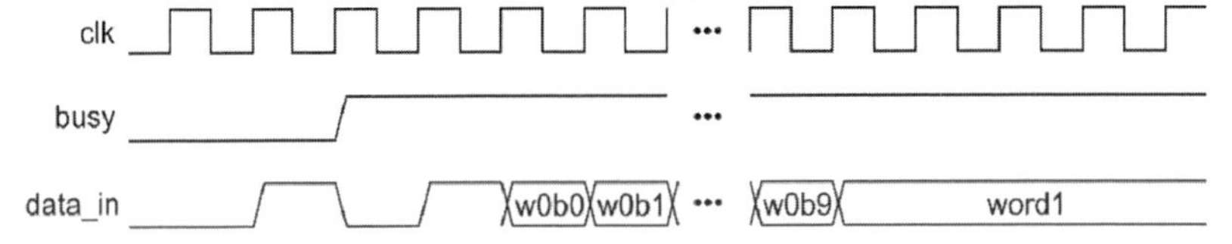

Figure 4.1.3 Time diagram of data_in, ASIC v1.

If the module is not busy and receives a rising edge on *start*, it goes into stimulation mode. In this mode the stimulator delivers only a single pulse, it releases the information stored in the memory through its *tdata* output, one word per clock pulse, starting with the word0 and ending with the word127. If start is cleared to low it goes into waiting mode. If *start* remains high, once the word127 is reached the cycle begins again with the word0 without latency clock pulse, this mode delivers the information as a continuous curve.

The 8 LSB´s that contain the magnitude of the pulse (*mdata[7..0]*) are taken by a code converter, this information will be available at *tdata* in thermometer code in a bus of 255 bits. The Most Significant Bit (MSB) that contains the sign of the pulse (*mdata[8]*) is taken by the "*H_Bridge Controller*" submodule. The outputs *sel_+* and *sel_-* stay low until the stimulation begins, during the stimulation only one of them goes high in order to drive out the required pulse at the output, either positive or negative. When the stimulation finishes, both of the mentioned outputs go to low in order to disconnect the pads of the output stage.

Stimulator, Voltage Bias Circuit

In order to keep the bias voltage stable versus changes of voltage from power supply, we implemented the Beta-multiplier circuit, shown in Figure 4.1.4, from [109].

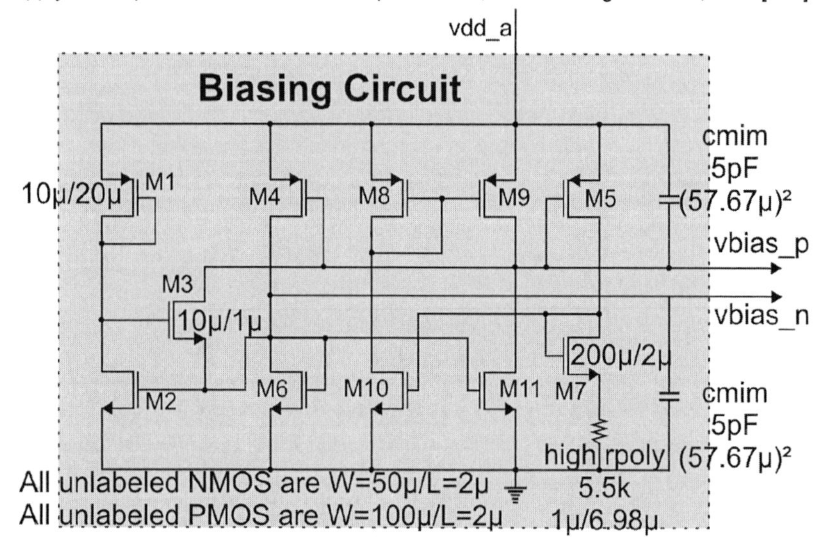

Figure 4.1.4 Biasing circuit, ASIC v1.

This circuit consists of three sub-circuits: a beta multiplier formed by the transistors: M4, M5, M6 and M7; a differential amplifier formed by the transistors: M8, M9, M10 and M11; and a start-up circuit formed by the transistors M1, M2 and M3. The condensators are the compensation capacitors, they are used to stabilize the reference current. The resistor is used to generate a reference current in the beta multiplier circuit.

In order to feed the differential amplifier, the drain's voltage of M6 and M7 should be similar. However, the V_{GS} voltage of M6 is higher than the V_{GS} of M7, because of the resistor. To force the same current through M6 and M7, the equation $V_{GS,M6} = V_{GS,M7\ sat} + I_{REF} \times R$ must be valid, which can only be valid if VGS of M6 is higher than VGS of M7. To ensure that, a larger value of ß is used in M7, i.e. the ß of M6 is multiplied in M7, so that less gate-source voltage is needed to conduct I_{REF}, this is done by using a larger width in M7. Since of that, the circuit is called beta multiplier.

The transistors M1, M2 and M3 form a start-up circuit for the beta multiplier. This start-up circuit is required when zero current flows in the circuit. Thus, the transistors M4, M5, M6 and M7 are switched OFF, which is an undesirable state of operation for the beta multiplier. At this point of time the gate voltage of transistors M4, M5 are close to VDD and gate voltage of transistors M6, M7 are close to ground, where VDD is the supply voltage. During this undesirable state, the transistor M2 is turned OFF because its gate is connected to the gate of M6 which is close to ground. The gate of M3 is somewhere in between VDD and Vthp, which is the threshold voltage of PMOS transistors and as a result of this the NMOS transistor M3 turns ON and leaks the current from the gate of M4, M5 to the gate of M6. This process turns the transistor M6 and M2 ON. Thus, it turns the transistor M3 OFF. This process aids the circuit to come out of the undesirable state. During normal state of operation, transistors M1, M2 and M3 which form the start-up circuit must not affect the beta multiplier.

Short-channel devices have low output resistance and as result of this, the drain current changes significantly with changes in drain-to-source voltage. With the usage of the differential amplifier, we reduce the sensitivity of variations in the drain-to-source with changes in VDD.

During the normal operation of the circuit, the differential amplifier is comparing the drain voltages of M6 and M7, if they are not equal, an output is generated which increases or decreases the V_{GS} of M4 and M5. As result, the drain voltages of M6 and M7 are equalized. The power supply rejection of circuit is improved with this feedback loop. The constant bias voltage generated by this circuit can be used to feed the gate of a transistor in a current mirror to replicate a current source independent of the variations in supply voltage.

Figure 4.1.5 shows the simulated outputs of the beta multiplier against a sweep of the supply voltage. Vbias P and Vbias N are constant between 0.44 V and 1.4 V, beyond this point both nodes start to increase slowly. Vbias P was measured as differential net against VDD.

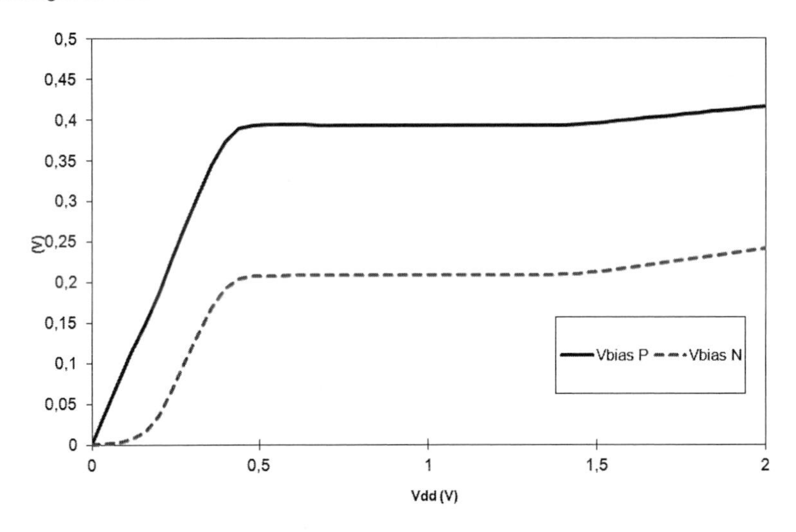

Figure 4.1.5 Beta multiplier outputs vs. Vdd, ASIC v1.

Figure 4.1.6 shows the simulated power supply rejection of the beta multiplier. The ratio for Vbias P is -60.70 dB from 1 Hz to 200 kHz. The ratio for Vbias N is -72.60 dB from 1Hz to 40 kHz.

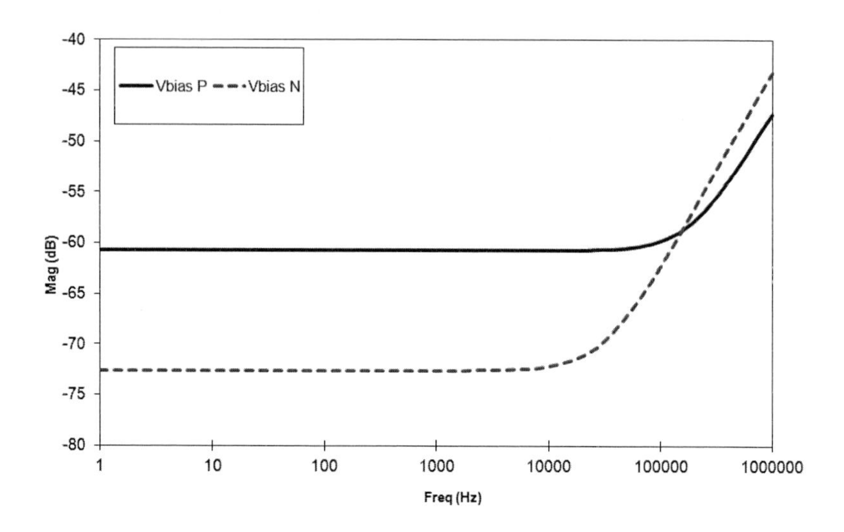

Figure 4.1.6 Beta multiplier power supply rejection, ASIC v1.

Stimulator, DAC

The desired signal at the output is a current signal. Therefore, the chosen architecture for the DAC is current steering, since such structures perform the conversion directly from digital to analog current signal without a voltage stage. Thus the design is simpler with less conversion stages and lower power consumption.

Such architecture is composed of several current cells, all of them connected to a common node in order to sum the current that each cell drives. The binary weighted array is usually preferable for the current cells, because of its simplicity and reduction in the digital control logic. But with this structure is difficult to achieve good linearity. In an N-bit array, only N current sources are available with variable sizes of bit current. This can lead to a large Differential Nonlinearity (DNL) error and an increased dynamic error during major code transitions. When the new code signal value appears before or after the signal value of the previous code disappears, a glitch is seen. This phenomenon is due to the magnitude of a glitch which is proportional to the number of switches that are actually switching, The biggest glitches tend to occur at major code transitions which is the point where the MSB changes from low to high and all other bits change from high to low, and vice versa. In this case the current source for the MSB should to be 2^{N-1} times bigger than the LSB current source, it means, the MSB represents 2^{N-1} times more switches than the LSB.

These problems are reduced by implementing the unary array (thermometer decoded), which is formed by 2^N-1 current cells, each of them equally sized. The binary input code shall be converted to a thermometer code that turns the corresponding current sources off or on. Some of the disadvantages of a thermometer code array are the area and complexity, since for each cell is required a current source, a switch, and a decoding circuit. However, there are advantages for a thermometer coded DAC versus the binary type, since each level step is created by switching only a small current cell, even for the major transition at the binary input code. Then the DNL error and glitch problems are greatly reduced.

Figure 4.1.7 shows a simple current cell (a), composed of a MOS transistor. However, this kind of cell is known to show a non-constant output current, which increases as the voltage across the current source increases (b).

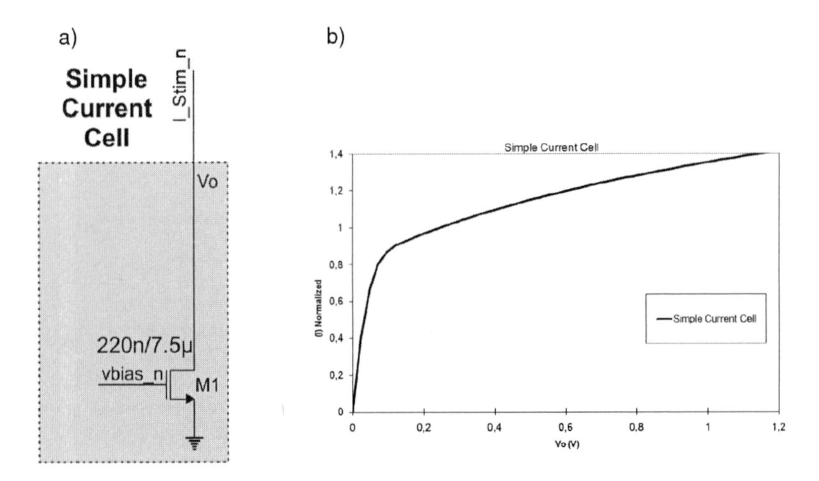

Figure 4.1.7 a) Simple current cell, b) operation of the current cell vs. a sweep of Vo, ASIC v1.

If we hold the drain-source voltage of the MOS constant, then the current does not vary. However, this requires a fixed V_o, which is our output node and will vary according the current's output. To make the MOS's drain voltage more constant, we will add circuitry in between M1 and Vo.

Since the basic current mirror has insufficient output resistance, cascoded current cells are used to improve its output resistance [110] [111]. Figure 4.1.8 shows a cascoded current cell (a), with its corresponding bias circuit. And its operation vs. different V_o levels (b).

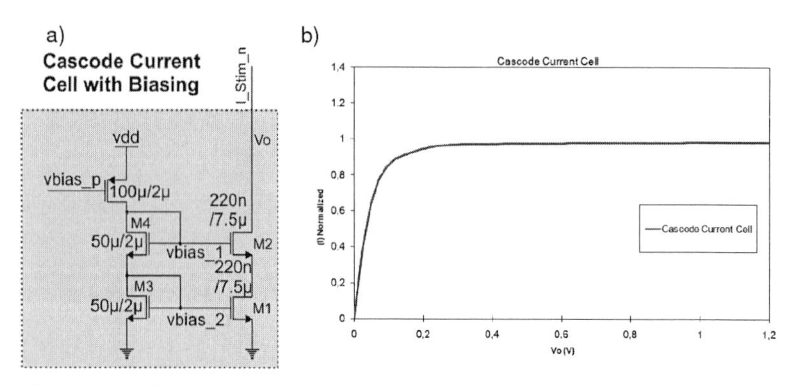

Figure 4.1.8 a) Cascode current cell, b) operation of the current cell vs. a sweep of Vo, ASIC v1.

It is important to realize that *I_stim* still determined by the gate-source voltage of M1. Changing the size of M2 simply changes the drain-source voltage of M1, affecting the matching of its drain current. The new schema attempts to hold the drain source voltage of M1 more constant to increase the current mirror's output resistance (make *I_stim* less sensitive to changes in V_o).

The voltage on the gate of M2 is $2V_{GS} = 2(V_{DS,sat} + V_{THN})$. The voltage on the drain of M1, assuming that it is operating in the saturation region, is V_{GS}. The minimum voltage across the current source is then: $V_{o,min}=V_{DS,sat2} + V_{GS} = 2V_{DS,sat} + V_{THN}$.

To keep both M1 and M2 operating in the saturation region, we need only a $V_{DS,sat}$ across each one. By biasing the cascode structure with a current mirror as shown if Figure 4.1.9, it is possible to bring the drain of M2, V_o, to the minimum possible voltage that keeps M1/M2 in saturation, this voltage is $2V_{DS,sat}$; The name of this structure is Wide-Swing Cascode, from [109]. Here the gate voltage of M1 will be $V_{DS,sat} + V_{THN}$ while the gate voltage of M2 will be $2V_{DS,sat} + V_{THN}$ and their drain voltages could be $V_{DS,sat}$ and more or equal than $2V_{DS,sat}$, respectively.

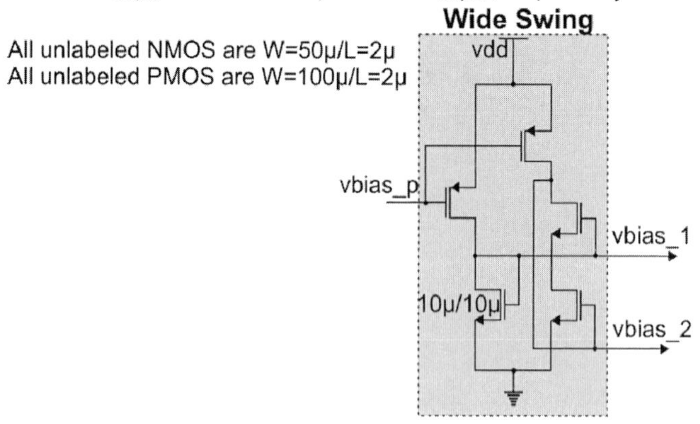

Figure 4.1.9 Wide swing biasing circuit for the cascode current cell, ASIC v1.

Figure 4.1.10 shows the operation of the cascode current cell with the basic biasing circuit and with the wide swing structure. The current was normalized to 1. For the simple circuit, it is shown that *I_stim* remains almost flat from 1.2 V to 250 mV where M2 start to triodes, and then the curve decreases slowly until 100 mV where M1 triodes, from this point the current decreases drastically. For the wide swing structure, the *I_stim* remains almost flat from 1.2 V to 100 mV which is the minimum voltage to keep M1 and M2 in saturation. In this case, the slope where M2 starts to triodes was smoothed.

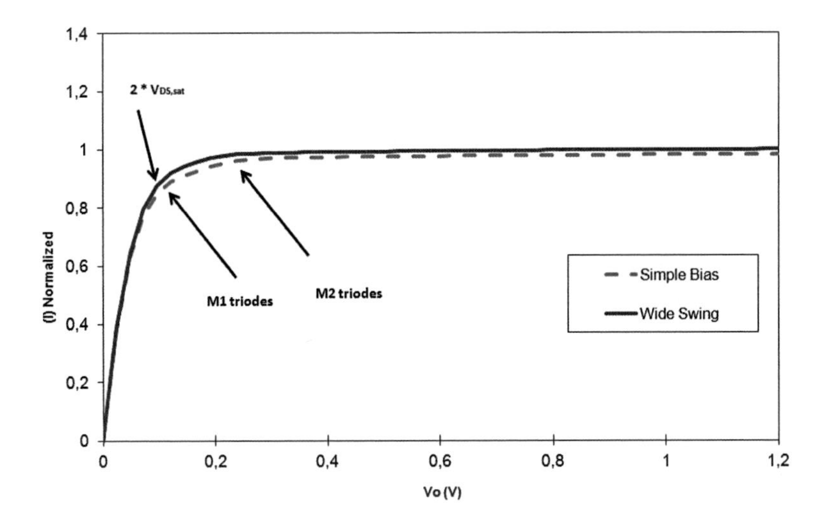

Figure 4.1.10 Operation of the cascode current cell with the simple bias circuit and with wide swing structure, ASIC v1.

Then in summary, the current sources that drive the output current shall present a large output resistance in order to be able to drive higher load resistances and also to present higher voltage dynamic range. It is important to bias both M1 and M2 in saturation and to hold the drain source voltage of M1 constant in order to improve the linearity and increase the current mirror's output resistance. This could be achieved by implementing a current cell formed by a cascode and with a wide swing biasing structure.

The current cell used in the ASIC is illustrated in Figure 4.1.11.

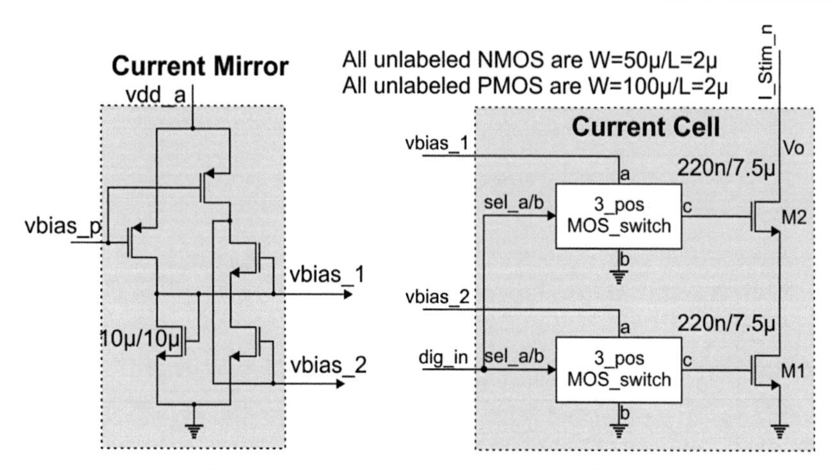

Figure 4.1.11 Current mirror and current cell, ASIC v1.

The switches that interrupt the current's flow are located normally between the load and the current cell, we chose to put them at the transistor gates in order to minimize the number of transistors in the branch to avoid voltage drops.

Figure 4.1.12 is a simulation of the cascode structure implemented in the ASIC, where a voltage source was connected at the node *Vo*, and the voltage was swept from 0 to 1.2 V. The current was normalized to 1. There it is shown that *I_stim* remains almost flat from 1.2 V to 100 mV, from this point the current decreases drastically. The voltage curves show how the drain voltage of M1 remains stable, allowing to keep constant the output current. Through this simulation it can be seen that this circuit could be useful for driving current signals with a maximum voltage around 1.1 V.

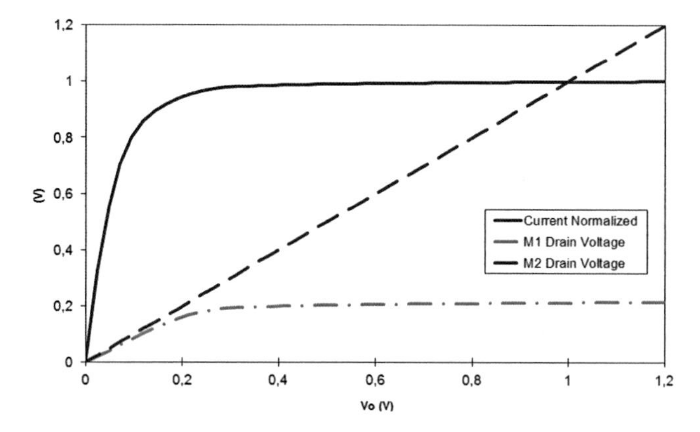

Figure 4.1.12 Simulation of the cascode structure versus a sweep voltage, ASIC v1.

Another advantage of these structures is the low power consumption because the current is flowing through the branch only when the current cell is turned on, and this current is the same that flows through the load.

Stimulator, H-Bridge

The output stage is driven by a MOS H-Bridge, Figure 4.1.13. This circuit makes possible the change of polarity at the output by activating the transistors connected either to *sel_+* or *sel_-* . By simply deactivating all the transistors in the H-Bridge, it is also possible to let the outputs *e1* and *e2* floating so they can serve as electrodes for biosignal acquisition.

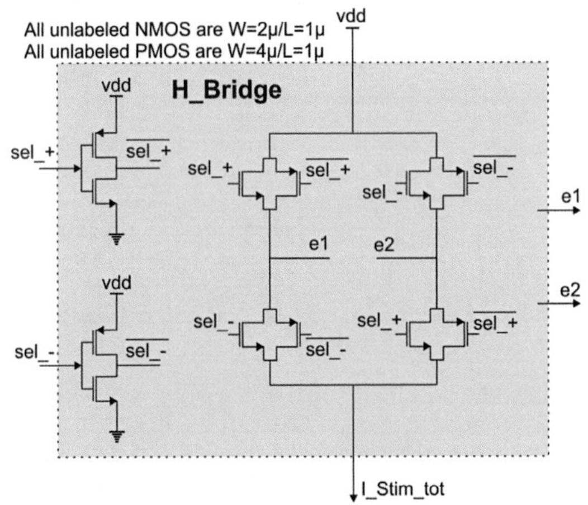

Figure 4.1.13 H-Bridge, ASIC v1.

The function of the circuit is determined by the following table:

Table 4.1.1 H-Bridge function, ASIC v1.

sel_+	sel_-	Function
0	0	Pins disconnected
0	1	Negative stimulation (from e2 to e1)
1	0	Positive stimulation (from e1 to e2)
1	1	FORBIDDEN

Analog Channel for Biosignals

The analog channel is used to amplify the bio-signals at its input. It delivers an amplified analog signal at the output. Figure 4.1.14 shows the modules of the channel which are: Operational Transconductance Amplifier (OTA), used at the input. The OTA is ac-coupled in order to cancel dc-offset of the input signal; Third order analog low-pass filter made out of three OTAs; An OTA in order to drive the next component; A Rail-to-Rail post amplifier.

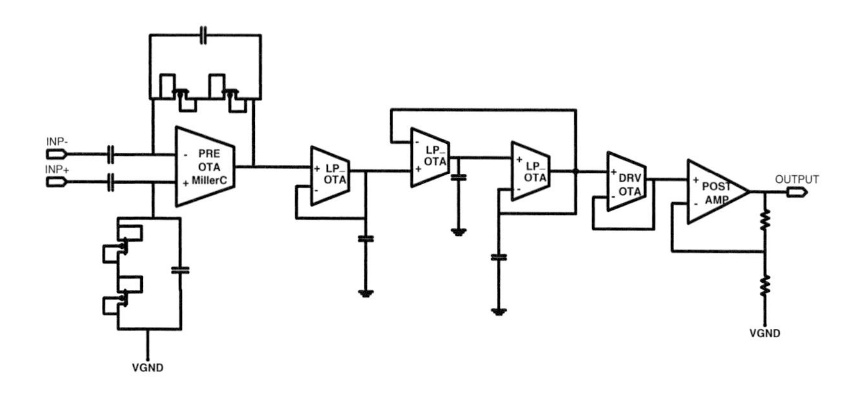

Figure 4.1.14 Structure of the analog channel for biosignal acquisition, ASIC v1.

4.1.2. Experimental Results

The layout of the fabricated ASIC is shown in Figure 4.1.15, where each module is identified. The ASIC was fabricated in the CMOS 130 nm / 1.2 V process. The design of the analog stage was done at transistor level and the digital control was designed in Very high speed integrated circuit Hardware Description Language (VHDL) and synthesized. The dimensions of the ASIC are 0.96 mm x 0.725 mm, the total area is 0.696 mm^2.The circuit is supplied with 1.2 V for the core and 3.3 V for the digital pads. For packing information refer to appendix C.

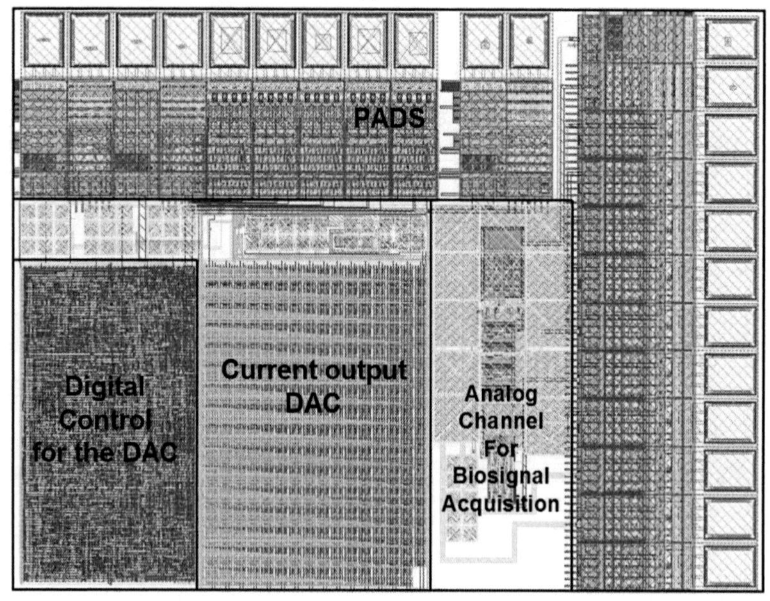

Figure 4.1.15 Floorplan of the ASIC v1.

Stimulator

The area for the stimulator is 0.257 mm^2, from which 0.095 mm^2 is occupied by the digital control, 0.146 mm^2 by the DAC including the current cells and the h-bridge and 0.016 mm^2 by the voltage biasing circuit. By setting a stimulator clock of 125 kHz, an output sinusoidal signal of 976 Hz at maximum amplitude, the power consumption of the stimulator was found to be 22.93 µW in standby mode, 120.83 µW in stimulating mode without load resistor and 128.12 µW with a 10 kΩ load. Simulations showed the power consumption for the "*Voltage Bias Circuit*" to be around 80 µW.

The transfer function for the positive pulse with different load resistors is shown in Figure 4.1.16. The maximum output current is ±9.8 µA. For loads requiring less than the maximum voltage swing, ±1.097 V, the negative pulse (not shown) presents a mismatch around 0.3%. There, it is also possible to see the DNL and INL, its average is 0.13 and 0.40 LSB, respectively, with maximum values of 0.47 and 1.05 LSB.

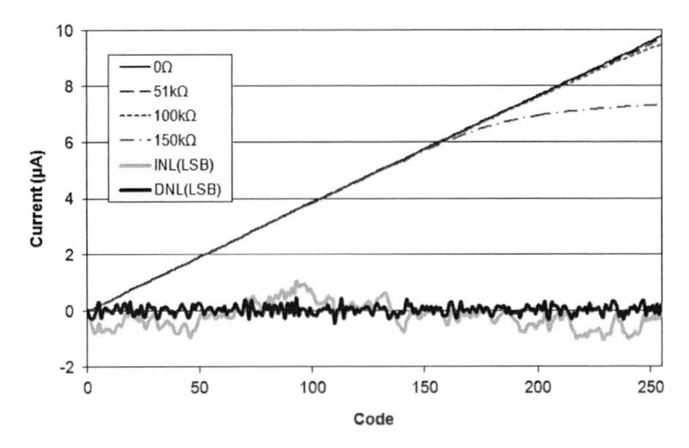

Figure 4.1.16 Transfer function of the stimulator for positive output with different load resistors and its DNL and INL, ASIC v1.

Figure 4.1.17 shows the Fast Fourier Transform (FFT) for the same setup with a load resistor of 10 kΩ. The FFT bandwidth was limited to 10 kHz, because of the range of interest for the stimulation signals. The spurious free dynamic range SFDR was found to be 50.2 dB, the DC level was neglected.

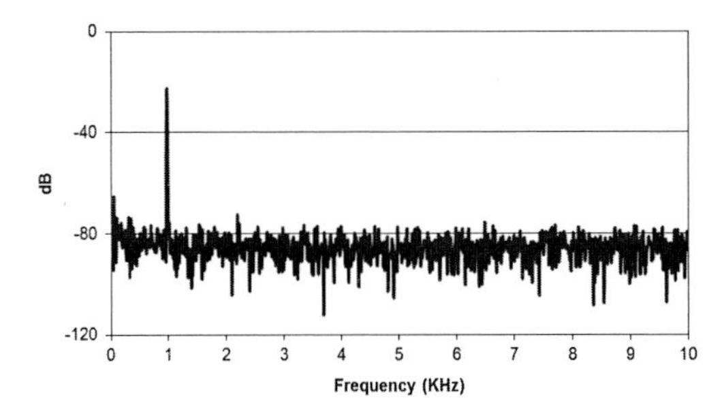

Figure 4.1.17 Measurement of the SFDR through the FFT of a sinusoidal, ASIC v1.

Three different stimulation waveforms were injected into a 10 kΩ resistive load and their respective voltage drops are shown in Figure 4.1.18.

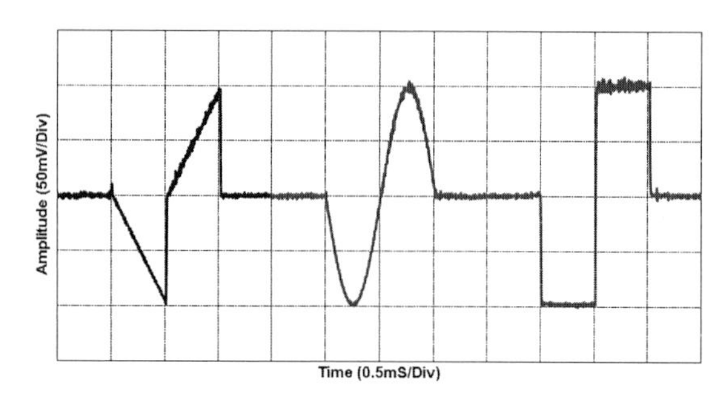

Figure 4.1.18 The voltage drop across a 10 kΩ resistive load for different waveforms, ASIC v1.

Analog Channel

The area of the analog input channel is 0.091 mm^2, its power consumption was found to be 140 μW. Using a signal analyzer, with an output level of 1 mV, and a frequency range from 10 Hz to 21 kHz, the gain for the whole channel was found to be 52.2 dB, as shown in Figure 4.1.19. The upper value of its bandwidth was measured at 1.13 kHz, and through simulations the lower value was found at 0.5 Hz. The output noise level measured is 10 μV_{rms}.

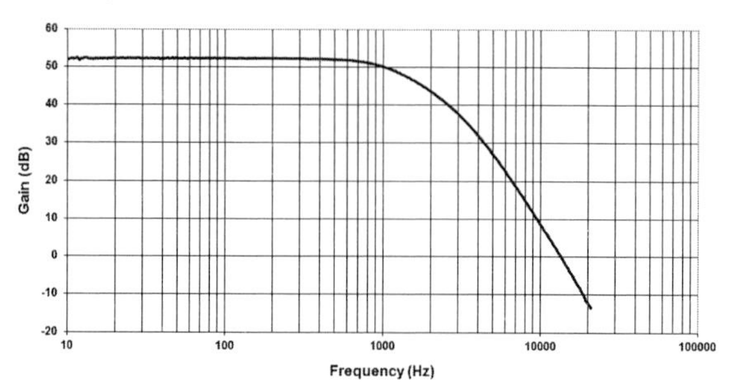

Figure 4.1.19 Transfer curve of the analog channel, ASIC v1.

Table 4.1.2 Measured specifications of the ASIC v1.

Technology	130 nm CMOS
Area	0.696 mm^2 (0.960 x 0.725 mm^2)
Supply Voltage	1.2 / 3.3 V
Stimulator	
Area	0.257 mm^2
Power Consumption	22.93 μW[a] / 128.12μW[b]
Resolution	8 / 9 Bits
DNL/INL	0.47 / 1.05 LSB
SFDR / Input Freq.	50.2 dB / 976 Hz
Max. Output	±1.097 V / ±9.8 μA
LSB	38.4 nA
Amplifier	
Area	0.091 mm^2
Power Consumption	140 μW
Gain	52.2 dB
Bandwidth	0.5 Hz – 1.13 kHz
Noise	10 μVrms

a. In standby mode, b. In stimulating mode, 976 Hz sinusoidal signal with a 10 kΩ load.

4.1.3. Discussion

During stimulation, approximately 62% of the power is consumed by the "*Voltage Bias Circuit*". Thus, the DAC architecture presented could be used in an array of several DAC´s having only one "*Voltage Bias Circuit*" and even only one "*Digital Control Unit*" without a significant increase to the overall power consumption. The same principle applies for saving silicon area, the actual area occupied by the pads is around 50% of the total area; the DAC itself occupies 21%, thus it could be possible to add more than one DAC by sharing the modules such "*Voltage Bias Circuit*" and by using the same number of pads to program the circuit in a serial way.

According to simulations performed by [77], the maximum current output of 9.8 µA is enough to activate motorneurons around 100 µm far away from the electrode. This current could be injected, for example, in Iridium Oxide electrodes with a diameter of 15 µm, which present an impedance of 113.6 kΩ at 1 kHz [56], or even in smaller electrodes coated with PEDOT, in order to increase the selectivity of the stimulation, see chapter 2. In case of Electrical Muscle Stimulation (EMS), or other applications requiring higher currents and voltages, an output amplifier could be attached, which could be implemented in High-Voltage-Laterally-Diffused-Metal-Oxide-Semiconductor (HVLDMOS) process, by using such transistors with 250nm technology it is possible to drive up to 80 V. Due to power consumption issues it is a better option to implement the stimulator in low voltage process and to attach a high voltage amplifier, than implement the whole system in a high voltage process, as was shown by [112].

It is possible to migrate the design to thick oxide transistors of the same technology, its breakdown voltage is 5 V. Thus, it is possible to supply the system with 3.3 V. Other concerns will be to decrease the required silicon area, to simplify the digital control and to reduce the complexity of the layout because of the amount of connections for controlling the unary array.

A hybrid architecture of the current steering DAC could overcome the negative aspects of the unary array and binary weight array. Since binary array DACs have problems associated with the MSB, it is suitable to be used on the LSB side of the DAC to handle the first few bits. For higher order bits a unary array can be used because this architecture can reduce the glitch effect introduced due to MSB switching. Thus, it could be possible to design a smaller stimulator because of the reduction of number of current cells, and also to increase the maximal current because of the operation voltage of the transistors.

4.2. *ASIC Version 2*

In the previous section we showed the unary current steering DAC architecture as the optimal option for a neurostimulator. This architecture was chosen because of the necessity for current at the output rather voltage, also because of its low power consumption, its capability of sharing some common stages for a multiarray stimulator and also for its simplicity. However, the voltage compliance was 1.2 V for the previous design, although in many applications the required voltage is higher because of the impedance of the electrodes.

In this section, are shown the results of an ASIC that we published in [107]. The ASIC has four stimulator channels on-chip formed by a hybrid architecture DAC implemented with thick oxide transistors. The design is implemented in 130 nm CMOS technology.

The hybrid architecture of current steering DAC is suitable to the application, because of its specifications: the ability to convert several waveforms directly from digital to analog current signals, low power consumption, small chip area requirement, the capability of sharing common stages and also for its simplicity.

It is also shown how such design can handle up to 3.3 V, which allows to drive smaller electrodes in order to deliver the necessary current, it is also shown how the hybrid architecture can keep the benefits of the unary current steering architecture in combination with smaller chip area.

4.2.1. ASIC Description

The first step was to migrate the design to high voltage transistors (thick oxide transistors) of the same technology, IHP 130 nm. Thus, it is possible to supply the system with 3.3 V. Another concern was to decrease the required silicon area, to simplify the digital control and simultaneously to reduce the complexity of the layout because of the amount of connections for controlling the unary array.

The block diagram of the system is shown in Figure 4.2.1. There is only one voltage bias circuit for the four DAC. Thus, it is possible to save area and to decrease the power consumption.

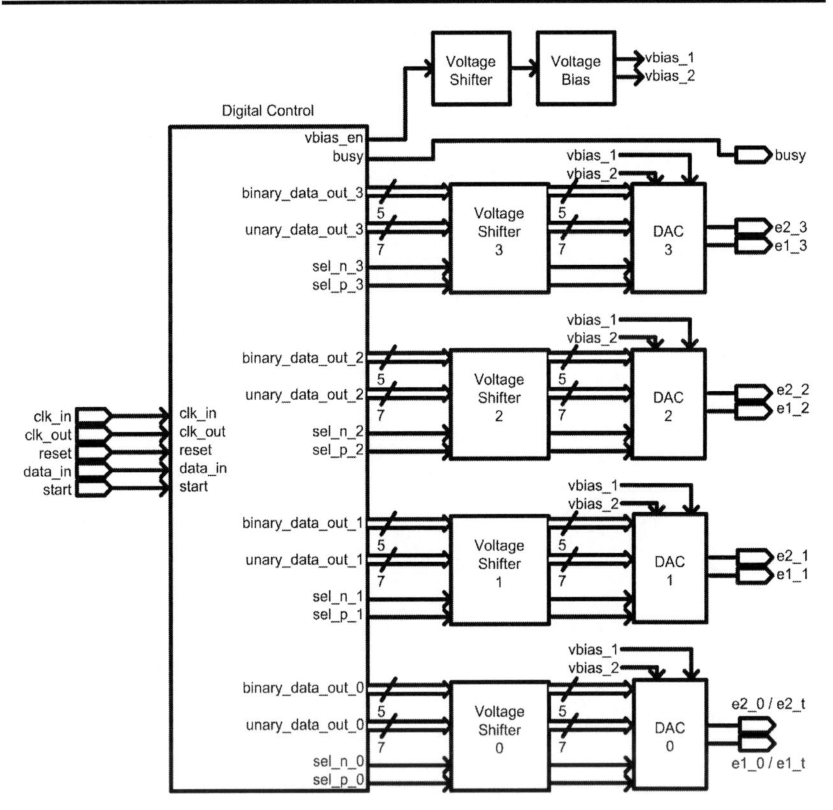

Figure 4.2.1 Diagram of the system, ASIC v2.

Digital Control

The digital control receives the information in serial way, it stores the data in four different memories (9 bits x 128), one for each channel, in this manner it is possible to stimulate with different amplitudes, pulse widths and waveforms on every channel. The memories are dual port which enables the programming of each individual channel while the stimulation is running.

It has also a programmable clock divider for each stimulation channel in order to adjust the time between pulses.

The digital stage has also a format converter (3 bit binary code to 7 bit thermometer code) for enabling the unary array, which is controlled through thermometer code.

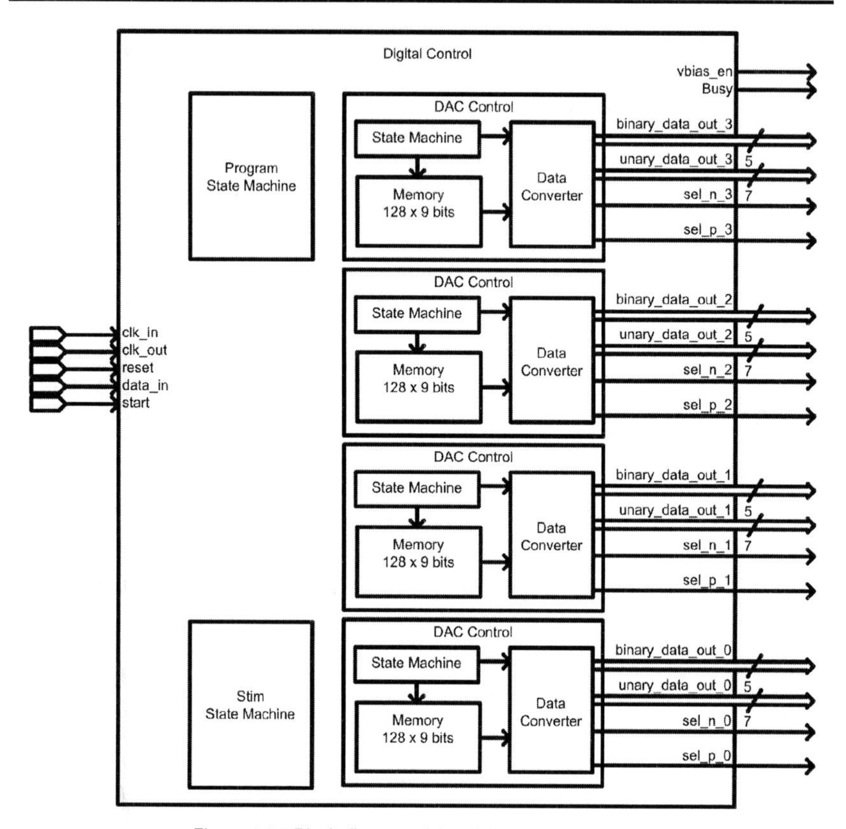

Figure 4.2.2 Block diagram of the digital control, ASIC v2.

All the digital logic reacts to rising edge clock, and its inputs and outputs are active high. It has two clock input lines, *clk_in* for the programming stage and *clk_out* for the stimulation stage. When this module is not busy, it is possible to program each individual channel by receiving a bit string with the value "*1001*" on its *data_in* line; after that, it receives two bits for choosing the desired channel, one bit to enable/disable the selected channel, nine bits for programming the time between pulses and then it receives 128 words of 9 bits each in consecutive order, Less Significant Bit (LSB) first as shown in Figure 4.2.3. The format of the words should be handed in Sign-Magnitude representation.

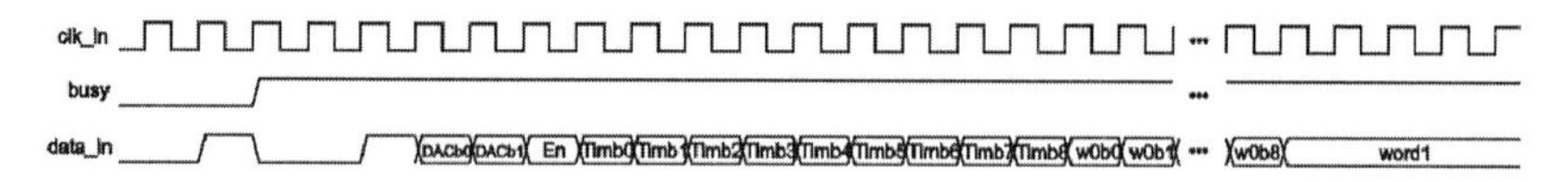

Figure 4.2.3 Time diagram of data_in, ASIC v2.

The nine bits for the time between pulses determine the number of clock pulses between a stimulation pulse and the next one.

The stimulator responds to the *clk_out*, if the module receives a rising edge on *start*, the four channels go into stimulation mode, except for the disabled through the enable register, their outputs remain isolated. In this mode the stimulator delivers only a single pulse, it releases the information stored in the memory, one word per clock pulse, starting with the word0 and ending with the word127. If start is cleared to low it goes into waiting mode. If *start* remains high, once the word127 is reached first the output pins are deactivated and they are isolated from the circuitry, the timer start to counting, once the timer is equal to the programmed timer the cycle begins again with the word0 without latency clock pulse, this mode delivers the information as a continuous curve. It is possible to reprogram each channel even during the stimulation.

Voltage Level Shifter

The digital stage is implement with thin oxide transistors and the analog with thick oxide transistors both of them working with different voltages, 1.2 and 3.3 V, respectively. Therefore, it is necessary to convert the control signals that come from the digital stage in order to drive the analog circuitry properly; each digital control line has its corresponding voltage level shifter.

There are two kinds of level shifter, inverter and no inverter, the former for converting the data of the DAC because it is implemented with P transistors that it requires inverted control, and also for controlling the H-Bridge, the further for enabling the voltage bias. In Figure 4.2.4 is possible to see the schematic for the no inverter voltage shifter. The inverter is implemented by exchanging the inputs of the HVNMOS transistors.

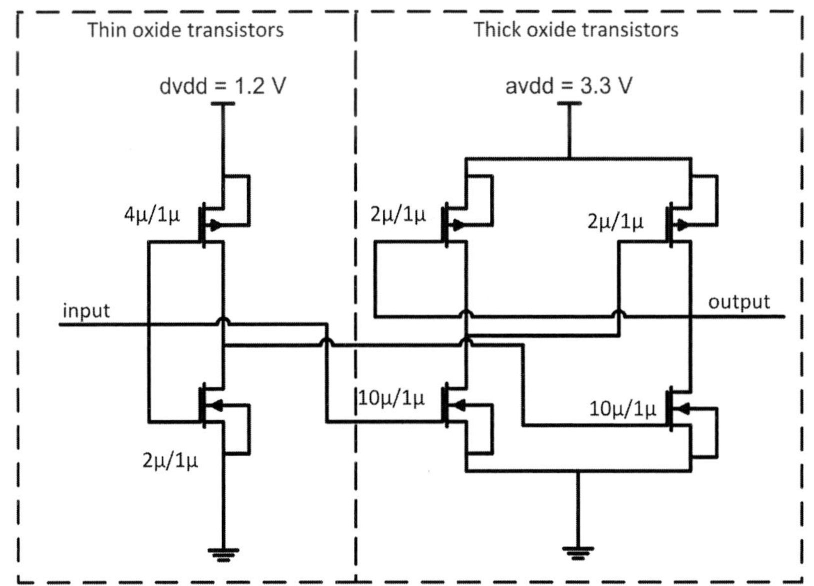

Figure 4.2.4 No inverter voltage level shifter, ASIC v2.

Figure 4.2.5 shows the typical curves of the level shifter with a binary current cell as load.

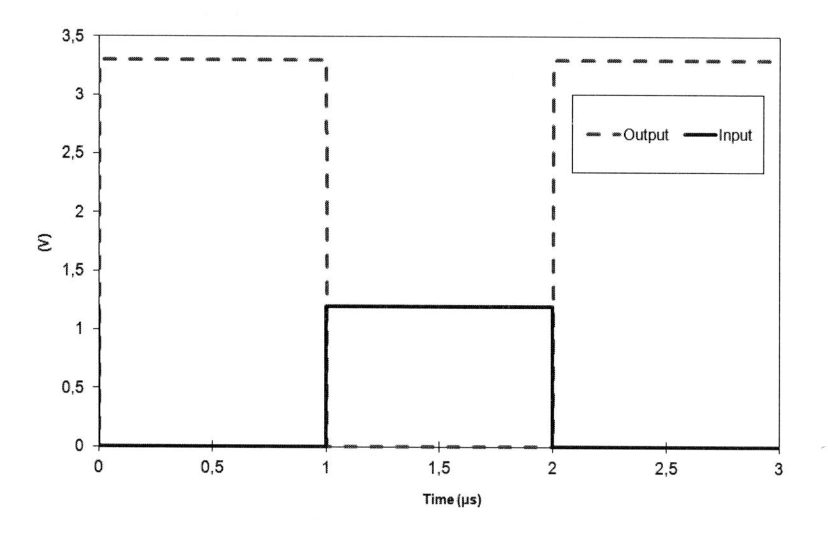

Figure 4.2.5 Input and output of the level shifter, ASIC v2.

Voltage Bias Circuit

In order to keep the bias voltage stable versus noise, changes of voltage from power supply, we implemented the same topology of the Beta-Multiplier circuit used in the ASIC version 1, Figure 4.1.4. We kept the same dimensions of the transistors, except M1, which now is 1µ/10µ. The wide swing current mirror is based on Figure 4.1.9. For this version the transistors used were thick oxide. The DAC is implemented with P transistors, that is why we need to mirror the wide swing current mirror. Figure 4.2.6 shows the diagram of the voltage bias circuit used in the ASIC version 2.

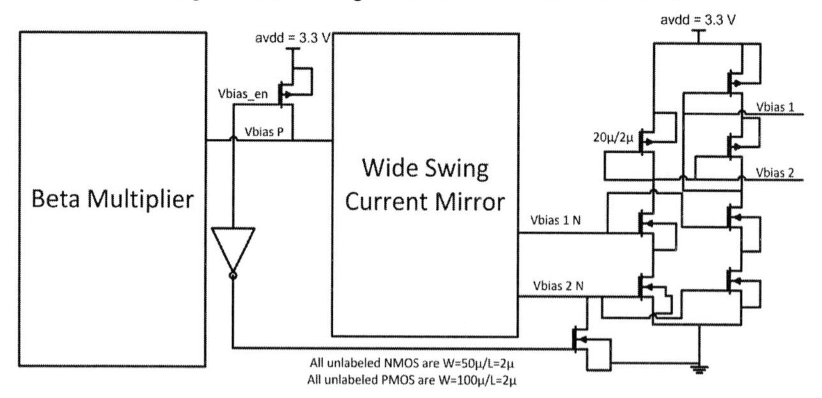

Figure 4.2.6 Wide swing current mirror, ASIC v2.

The voltage bias circuit was simulated with 4 DACs (next section), and the current output of each DAC is illustrated in Figure 4.2.7

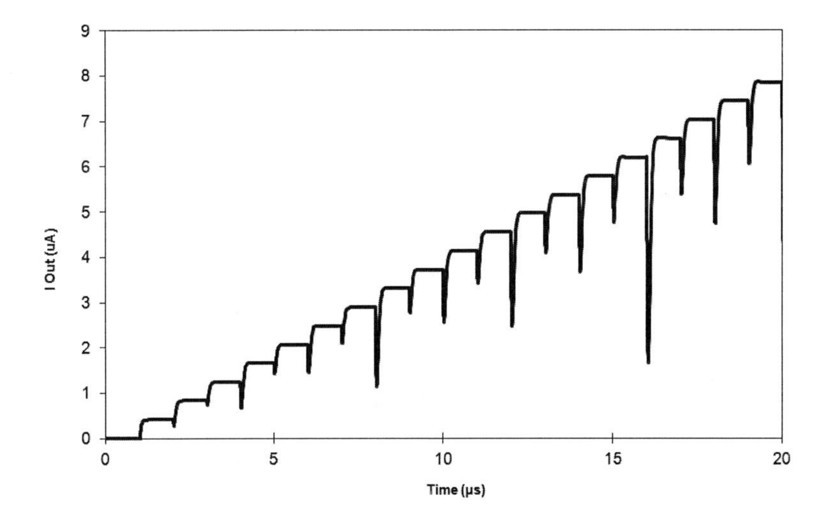

Figure 4.2.7 DAC´s output with single voltage biasing, ASIC v2.

The transition glitches are too high. This is because of the charge injection induced in the vbias 1 while the pass gates are switching. See Figure 4.2.8

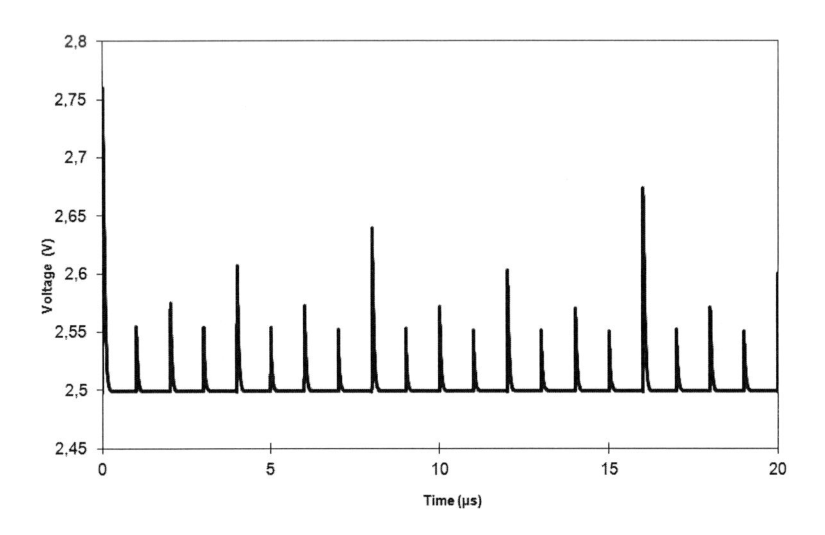

Figure 4.2.8 Vbias 1 with single circuit, ASIC v2.

To minimize the phenomena, it is necessary to reduce the output impedance of the voltage bias circuit. We mirrored the current with bigger transistors and added capacitors at the outputs to stabilize them. Figure 4.2.9

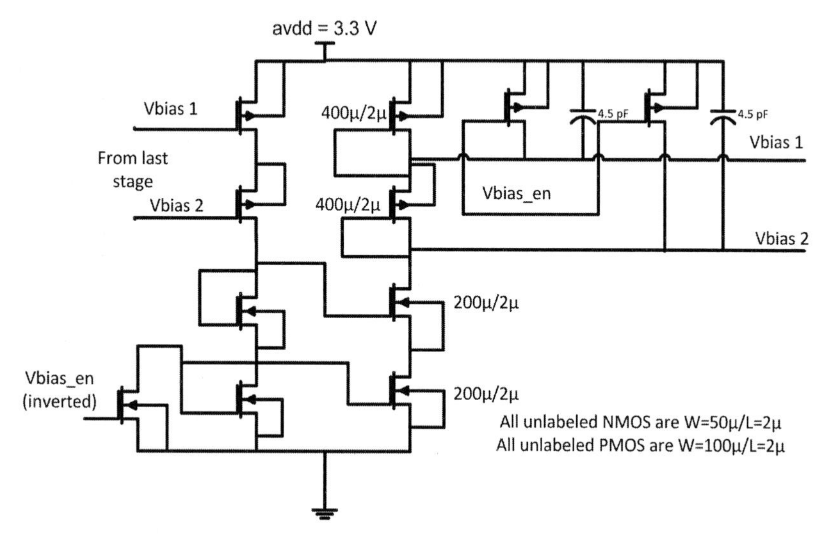

Figure 4.2.9 Extra current mirror, ASIC v2.

Figure 4.2.10 shows the outputs of the simple version voltage bias circuit and the new version with extra output transistors.

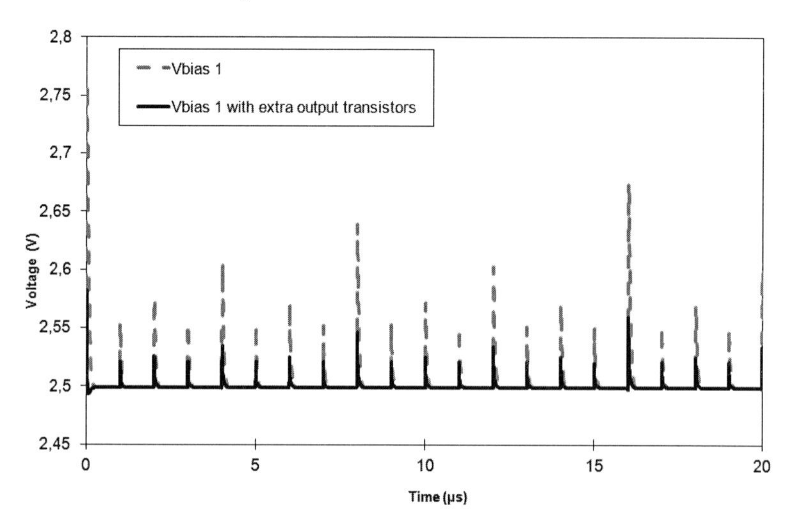

Figure 4.2.10 Vbias 1 with single circuit and with extra output transistors, ASIC v2.

The simulation of the DACs by using the two different versions of voltage bias circuits is illustrated in Figure 4.2.11.

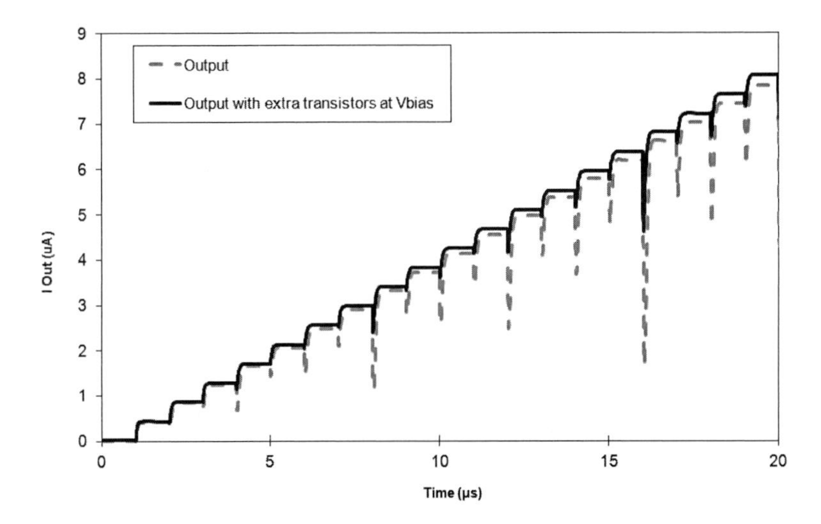

Figure 4.2.11 DAC´s output with single voltage biasing and with extra transistors, ASIC v2.

Figure 4.2.12 shows the simulated outputs of the beta multiplier against a sweep of the supply voltage. Vbias 1 and Vbias 2 are constant from 1.7 V.

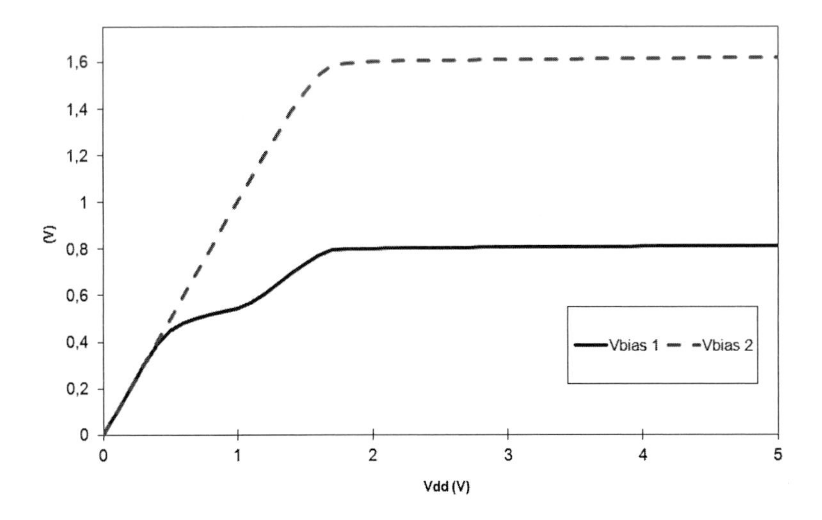

Figure 4.2.12 Beta multiplier outputs vs. Vdd, ASIC v2.

Figure 4.2.13 shows the simulated power supply rejection of the beta multiplier. The ratio for Vbias 1 is -46.33 dB from 1 Hz to 83 kHz. The ratio for Vbias 2 is -50 dB from 1Hz to 144 kHz.

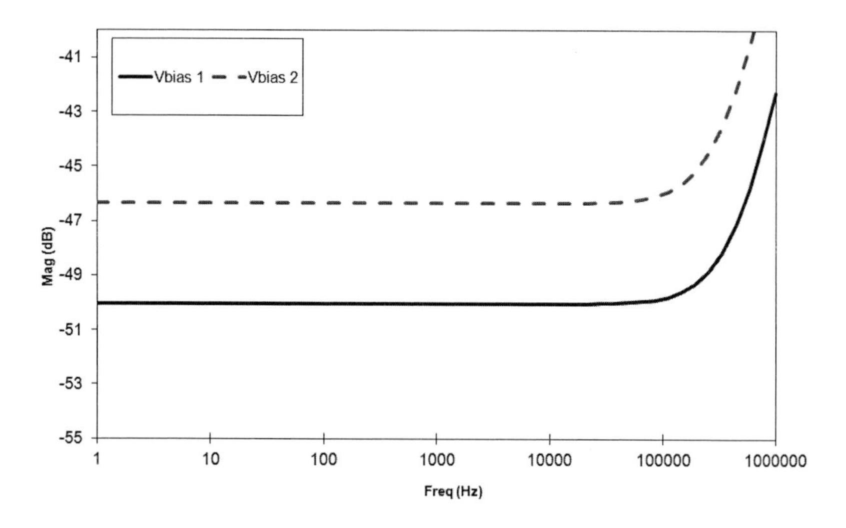

Figure 4.2.13 Beta multiplier power supply rejection, ASIC v2.

DAC

A hybrid architecture of the current steering DAC could overcome the negative aspects of the unary array and binary weight array. With the latter it is difficult to achieve good linearity. In an N-bit array, only N current sources are available with variable sizes of bit current, this can lead to a large DNL error, besides the glitch problems are inherent to the size of the transistors switching simultaneously. These problems are reduced by implementing the unary array which is formed by 2^N-1 equally sized current cells. However, some of the disadvantages of a unary array are the area and complexity, since for each cell a current source, a switch, and a decoding circuit are required, besides of the difficulty in performing the routing in the layout because of the large number of cells. Since binary array DACs have problems associated with the MSB, it is suitable to be used on the LSB side of the DAC to handle the first few bits. For higher order bits a unary array can be used because this architecture can reduce the glitch effect introduced due to MSB switching.

If we assume B1 as the number of desired bits for the unary array and B2 for the binary array, the required cells for the unary array are 2^{B1}-1 and the required cells for the binary array are equal to B2. The total number of cells is calculated with Equation 4.1 and the standard deviation [113] is calculated with Equation 4.2 and Equation 4.3.

Equation 4.1 $$S = (2^{B1} - 1) + (B2)$$

Equation 4.2 $$\sigma_{INL} \approx \sqrt{2^{(B-2)}}\sigma_E$$

Equation 4.3 $$\sigma_{DNL} \approx \sqrt{2^{(B2+1)} - 1}\sigma_E$$

B is the number of bits, 8 in this case, and σ_E is the error associated with element matching. Table 4.2.1 shows the INL and DNL standard deviation and the number of cells required for different combinations of a hybrid 8 bit DAC.

Table 4.2.1 Hybrid DAC architecture analysis, ASIC v2.

Unary (B1)	Binary (B2)	Total Cells (S)	σ_{INL}[LSB]	σ_{DNL}[LSB]
0	8	8	0.16	22.60 σ_E
1	7	8	0.16	15.96 σ_E
2	6	9	0.16	11.26 σ_E
3	5	12	0.16	7.93 σ_E
4	4	19	0.16	5.56 σ_E
5	3	34	0.16	3.87 σ_E
6	2	65	0.16	2.64 σ_E
7	1	128	0.16	1.73 σ_E
8	0	255	0.16	σ_E

Due to the equilibrium between number of cells and σ_{DNL}, we decided to use the combination of 3 unary bits and 5 binary bits. With an increase of number of cells the digital stage becomes bigger, as well as, the complexity of the layout because of the quantity of cells to control and to route. The DAC is formed by 12 current cells; 5 of them are binary weight scaled, where the smallest is equivalent to 1x LSB and the largest to 16x LSB; and 7 unary current cells each one represents 32x LSB, see Figure 4.2.14. All these values are represented in relation to the current mirror that drives the maximum output current, which is set to 100 µA.

The current cells should present a large output resistance in order to be able to drive higher load resistances and also to present higher voltage dynamic range. This is achieved by implementing a current cell formed by a folded cascode with PMOS transistors. It is important to bias both of the transistors in saturation and to hold constant the drain source voltage of the upper transistor in order to improve the linearity and increase the current mirror's output resistance. The transistors are biased with a Wide-Swing Cascode structure, as shown previously and explained in the section 4.1.1. Thus, it is possible to bring the drain of the lower transistor to the minimum possible voltage that keeps the two transistors in saturation, this voltage is $2V_{DS,sat}$.

Figure 4.2.15 shows the current in an unary current cell, 32x LSB, versus different voltages at the load from 0 to 5 V by supplying the design (including the voltage bias circuit) with three different voltages, 2.8, 3.3 and 3.8 V.

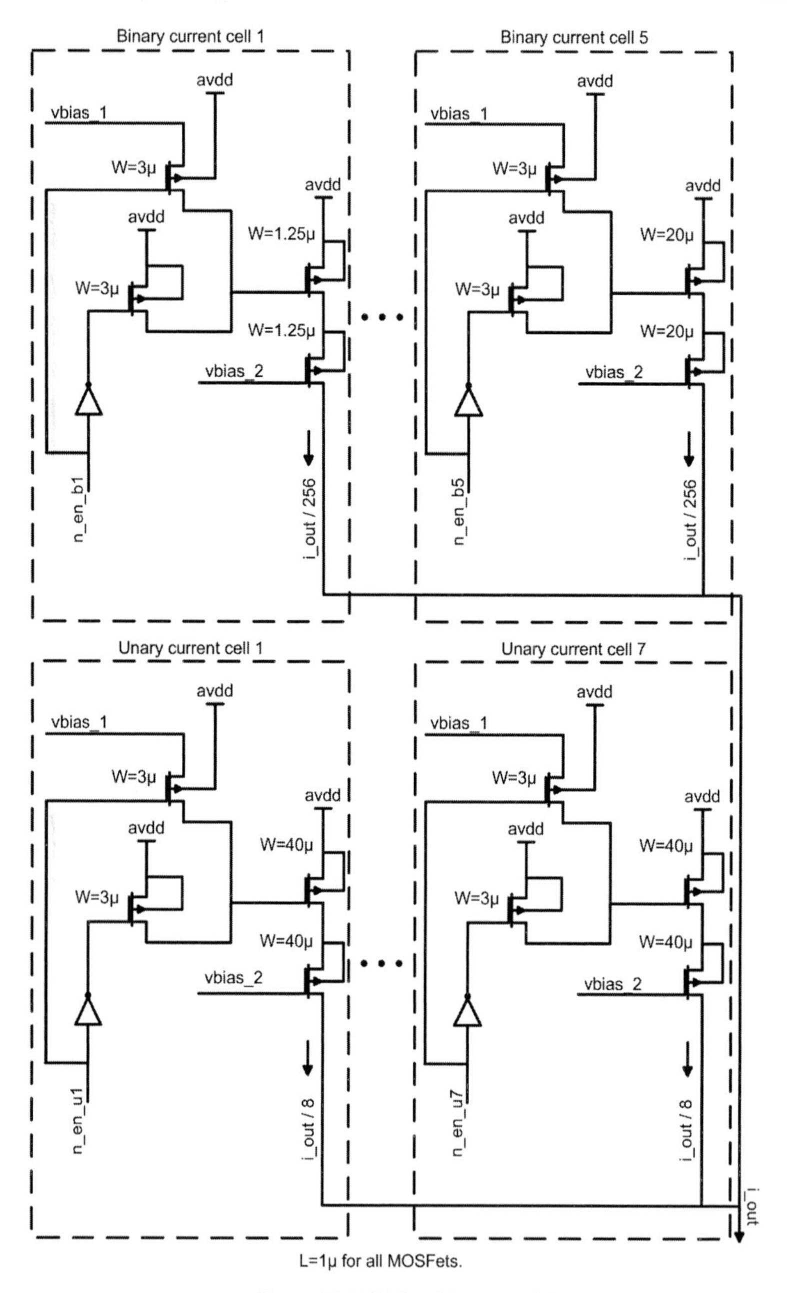

Figure 4.2.14 DAC architecture, ASIC v2.

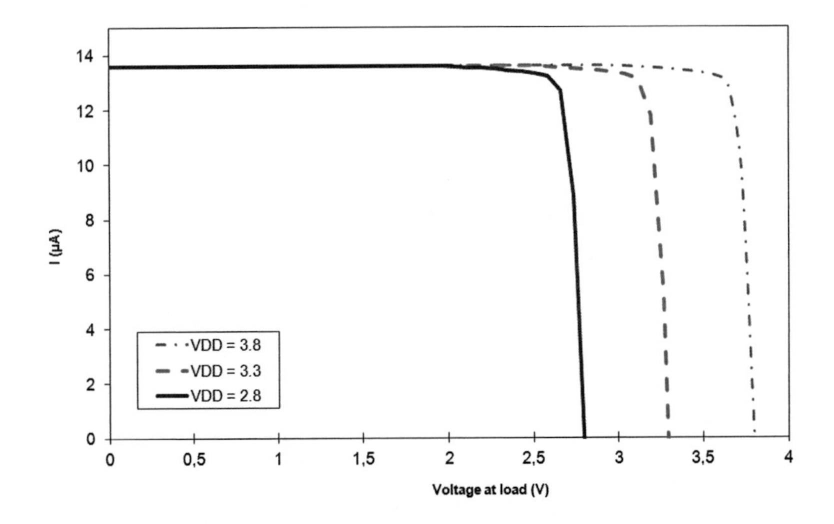

Figure 4.2.15 Simulated current in an unary cell vs. different voltages at load, ASIC v2.

H-Bridge

All the current cells are connected to a common node to sum the currents. At the output the current is driven for an H-Bridge, which enables the bipolar output by inverting the polarization of the output lines and makes it possible to isolate the output pins of the output stage for performing other tasks at the electrodes. Because of the bipolar output the DAC behaves like a 9 bit DAC, i.e. 256 different positive current levels, and also negative. The schematic corresponds to the same used in the ASIC v1, Figure 4.1.13. For this design, the transistors used were HVMOS transistors. The dimensions of the inverters are 10µ/1µ and 2.86µ/1µ, for the P and N transistors, respectively. And the dimensions of the H-Bridge are 30µ/1µ and 10µ/1µ, for the P and N types.

4.2.2. Experimental Results

The transistors were placed into the desired floorplan and routed using the Cadence layout suite. The resulting area is 1.058 mm^2 (850 µm x 1,245 µm) for the whole ASIC; 0.757 mm^2 (608 µm x 1,245 µm) for the four channel stimulator; 0.025 mm^2 for the voltage biasing circuit; 0.003 mm^2 for each bus voltage shifter; and 0.009 mm^2 for each DAC containing the 12 current cells and the h-bridge. Figure 4.2.16 shows a picture of the fabricated ASIC. Test structures and five transistors were implemented for test purposes, refer to appendix D.

By setting a stimulator clock of 125 kHz, an output sinusoidal signal of 976 Hz at maximum amplitude, and using a sub-femtoamp-meter (Keithley 630), the power consumption of the stimulator was found to be 56.81 µW in standby mode, 884.21 µW in stimulation mode without load, and 1.10 mW with a 10 kΩ load on only one channel. The power consumption for the voltage bias was estimated to be around 825 µW, by subtracting the power when the circuit is in standby from the power when it is stimulating without load.

Figure 4.2.17 shows the behavior of the circuit as function of the analog supply voltage. For these measurements, the test structures were used and programmed to drive out a dc current at maximum amplitude. The analog supply voltage was swept from 0 V to 3.5 V. The bias voltages were measured in relation to the supply voltage, i.e. the source of the output transistors. A $10\ k\Omega$ resistor was used as load at the output.

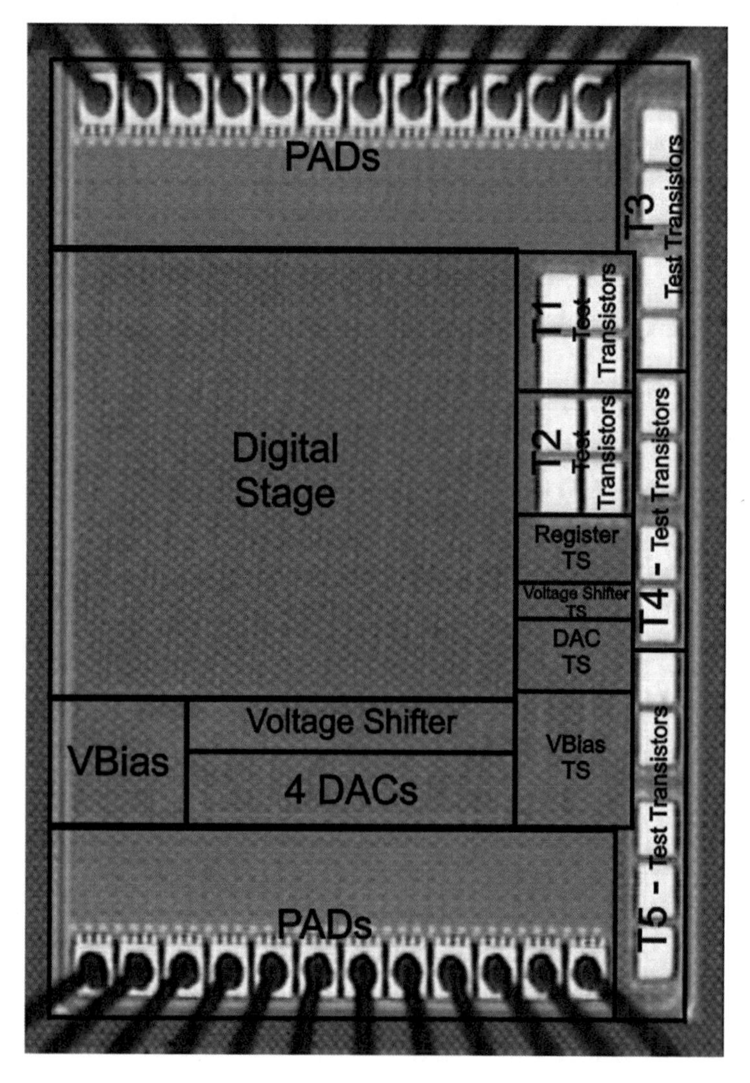

Figure 4.2.16 Micrograph of the ASIC v2 and its floorplan.

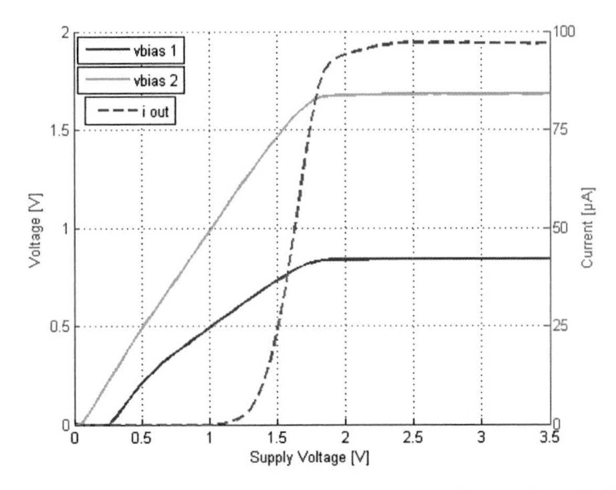

Figure 4.2.17 Voltage bias circuit output and current output vs change of supply, ASIC v2.

An *In Vitro* test was performed with 1000 µm^2 (35.7 µm of diameter) flexible gold electrodes. For the measurements, an electrode array was immersed in a plastic recipient containing saline solution (DeltaSelect GmbH), the distance between electrodes was 5 mm, the impedance between two electrodes (including the solution resistance) was found to be around 140kΩ at 1 kHz. It was injected into the electrodes symmetrical biphasic current pulses, without interphase and at maximum current amplitude, the waveforms were rectangular and sinusoidal, all of them had a width of 256 µs. The current and voltage forms were recorded through the computer by using an acquisition card (National Instrument DAQ M 6289 PCI). The plots are shown in Figure 4.2.18. The voltage on the electrode reaches the limit of the stimulator, see the discussion.

The experiment was repeated with the same electrodes coated with PEDOT/NaPSS galvanostatically deposited by injecting a charge of 180 mC/cm^2; they were prepared according to the chapter 2. Each electrode presents an individual impedance of around 12, 11, 7.5 and 6.5 kΩ at 0.5, 1, 5 and 10 kHz, respectively, which are within the range of interest for neurostimulation. The impedance between two electrodes (including the solution resistance) was found to be around 27 kΩ at 1 kHz.

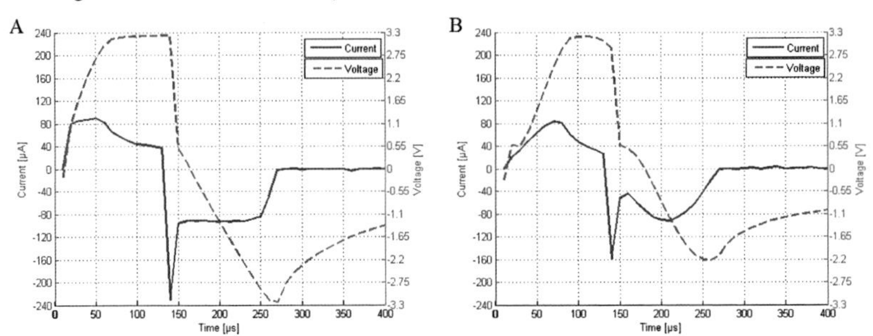

Figure 4.2.18 Current and voltage on gold microelectrodes, a) rectangular, b) sinusoidal, ASIC v2.

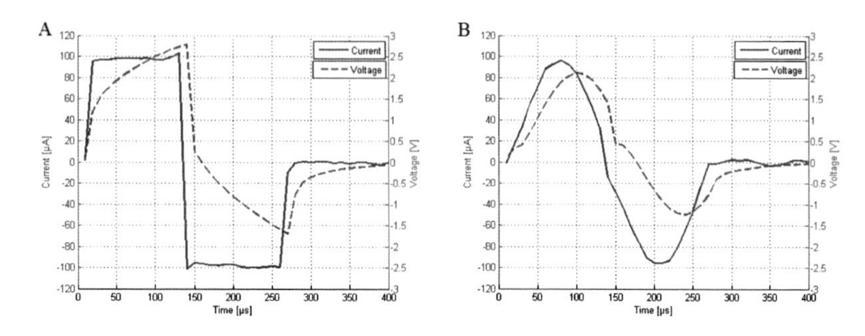

Figure 4.2.19 Current and voltage on PEDOT microelectrodes, a) rectangular, b) sinusoidal, ASIC v2.

By placing a variable resistor as load, the maximum output current was found to be ±97.10 µA and the maximum output voltage ±3.14 V. Table 4.2.2 summarizes the specifications of the hybrid architecture compared with the ASIC of the section 4.1.

Table 4.2.2 Measured specifications of the ASIC vs unary array.

	Unary Array	**Hybrid Architecture**
Technology	130 nm CMOS	130 nm CMOS
ASIC area	0.696 mm^2	0.757 mm^2 [a]
Supply voltage	1.2 / 3.3 V	1.2 / 3.3 V
Number of channels	1	4
	Voltage bias circuit	
Area	0.016 mm^2	0.025 mm^2
Power consumption	80 µW [b]	825 µW [b]
Power normalized to channel and voltage	66 µW/ChV [b]	62 µW/ChV [b]
	DAC	
Area	0.146 mm^2	0.012 mm^2 [c]
Resolution	8 / 9 Bits	8 / 9 Bits
Number of control lines (cells)	255	12
Max. output	±1.09 V / ±9.8 µA	±3.14 V / ±97.1 µA
LSB	38.4 nA	380.78 nA

a. Area for the four channels without the test structures, b. Estimated, c. Area with voltage shifter.

4.2.3. Discussion

In comparison with the unary array, the area occupied for the voltage bias circuit is larger because of the size increase of the output transistors, the increase was required due to the necessity of driving more channels. The power consumption of the voltage bias circuit was higher because of the larger output stage and also because of the changing of power supply, i.e. from 1.2 V to 3.3 V. For a design with higher number of stimulators it would be recommended to implement a voltage regulator at the output of the voltage bias, instead of increasing the size and power consumption of the transistors.

The power consumption of the DAC in standby is determined by the leakage current of the transistors, and during the stimulation it is equal to the charge delivered at the electrodes times the supply voltage, e.g. the RMS current value of the sinusoidal times 3.3 V, plus the power of the stimulation without load is the same as the power consumption by stimulating with a 10 kΩ load.

The voltage bias circuit is stable versus variations of supply voltage, as long as the voltage is above the minimum required for the circuit, around 1.85 V. The output current also remains stable. On one hand the circuit is stable versus supply voltage variations caused by noise at the supply line, on the other hand the power consumption could be decreased by reducing the supply voltage according to the output requirements. However, it is important to notice that the supply voltage should be big enough to permit the voltage drop at the output according to its impedance and the output transistors should still be in saturation.

The area requirement of the hybrid architecture DAC (including the voltage shifter) is twelve times lower than the unary array, and it is capable to drive out ten times more current and also higher voltages. The layout routing is simpler because of the reduction of number of control lines. Thus, the hybrid architecture facilitates the design of stimulators with higher number of channels, for example in [114] a retinal prosthesis composed of the retinal implant is presented, which contains the electrodes and a chip containing the circuitry, the former is a 5x6 mm^2 parylene substrate with an array of 1000 electrodes. Thus, by using the hybrid architecture, an array of 1000 DACs could be fitted and integrated together into the electrodes array, without the necessity of having two separate substrates. Besides, due to the increase of current and voltage output, with this hybrid architecture it is possible to stimulate neurons from a greater distance and to use smaller electrodes in order to achieve higher selectivity.

The current and voltage curves on gold electrodes show the importance of the electrode impedance for stimulation. When the impedance is high it is difficult to drive the electrodes with low voltage transistor technologies as it was seen in the curves, the voltage on the electrode reaches the limit of the stimulator. Due to it, the current could no goes until the programmed amplitude and the waveform is not the desired.

By covering the electrodes with the PEDOT the stimulator was able of driving microelectrodes small enough for performing invasive stimulation, e.g. in the *In Vivo* experiments in retina performed by [6], they used iridium oxide electrodes with a diameter of 100 μm, and they reported that visual sensations were elicited with currents as low as 3.2 μA. We have shown how our stimulator can deliver up to 97.1 μA on 35.7 μm diameter electrodes, which should present higher impedance than the 100 μm diameter electrodes because of the size. Thus, the stimulator is able to deliver the necessary amount of current in order to fire an action potential by using several electrodes.

4.3. ASIC Version 3

Small electrodes are required in invasive stimulation, and in some applications a large number of them is required; e. g. in muscular stimulation for restoring the mobility of extremities like the hand, simultaneous stimulation in different sites is required to stimulate different groups of nerves and perform natural movements; in neural stimulation for restoring the vision, i.e. retinal, visual nerve, lateral geniculate or cortical implants, between 600 to 1000 electrodes are required to give the perception of a basic vision enough to achieve simple tasks like navigating in a room or reading large-sized text [115].

The most basic tasks that a visually impaired person would like to perform, would be to recognize a person in front of them, to identify a door or read some text. In Figure 4.3.1 is possible to see three original pictures, a face; a door; and some text. Then, the pictures were treated as follows: they were changed to 8-bit grayscale color, color inverted, and edge detection filter was applied, the number of pixels was reduced (64, 256 and 1024), and a grid pattern was applied to have circular pixels on a black background. The result is used as simulation to predict the least amount of stimulation points required to give perception of a basic vision to a patient. With 64 pixels, the face and the door cannot be recognized, with 256 they can be poorly distinguished, and with 1024 is possible to distinguish a face or a door in the picture. Regarding the text, just in the 1024 configuration it is readable.

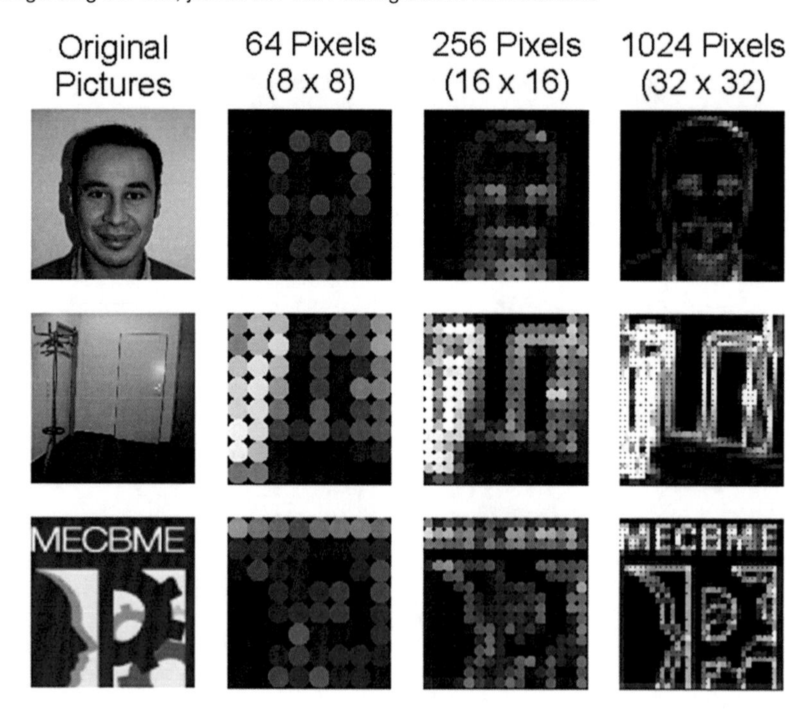

Figure 4.3.1 Simulation of the perceived image from patients with visual implants under different pixel configuration.

Charge balanced stimulation is also important especially in invasive stimulation, because an imbalanced stimulation may cause neural tissue damage [116].

In this section, we present a design that we published and described in [108]. The ASIC is an integrated neurostimulator formed by 64 Local Stimulation Units (LSUs) fabricated with 130nm CMOS transistor technology process, contained in a small area chip suitable for implantation. Each LSU is composed of an 8 bit DAC, digital stage and output multiplexer. The DAC offers the ability to deliver several waveforms in order to save energy. The digital stage on each LSU permits to refresh periodically the amplitude on every electrode for changing the intensity of individual stimulation. The output multiplexer allows the interaction of adjacent LSUs during the stimulation. The hybrid architecture of the ASIC version 2 was used for the DACs because of its advantages with respect to size and low power consumption.

The chip is fully scalable by connecting it in daisy chain configuration in order to increase the number of stimulation channels, for those applications that require a large amount of stimulation sites. All the individual channels on every chip will be synchronized in stimulation and refresh time.

The stimulation can be performed on all the LSUs simultaneously or each in a different timeslot to prevent cross electrode stimulation and thus avoid a charge imbalanced stimulation. Besides, the chip employs a schema which carries out the stimulation using exactly the same current sources during the anodic and cathodic part of the pulse, thus avoiding the residual charge due to the process mismatch.

The PEDOT electrodes show adequate characteristics for the application. However, the deposition of the polymer by galvanostatic deposition is performed by injecting a specific amount of current for a specific time, while the electrodes are immersed in a solution, see chapter 2. Sometimes due to the design of implants, it is not possible to have electrical access to the electrodes with a device different than the stimulator. Therefore, an on-chip module was implemented, capable to control the electropolymerization process of the electrodes.

4.3.1. ASIC Description

All the design was implemented in 130 nm CMOS transistor technology, the digital stage is working with 1.2 V thin oxide transistors and the analog part is working with 3.3 V thick oxide transistors, voltage converters are interfacing the digital and analog modules.

Figure 4.3.2 shows the block diagram of the design. Multiple chips can be controlled by only 6 digital lines. Four lines go in parallel to all the chips, and every chip has two inputs and its corresponding outputs to form a daisy chain configuration. By this it is possible to attach several ASICs, which allows increasing the number of stimulations sites.

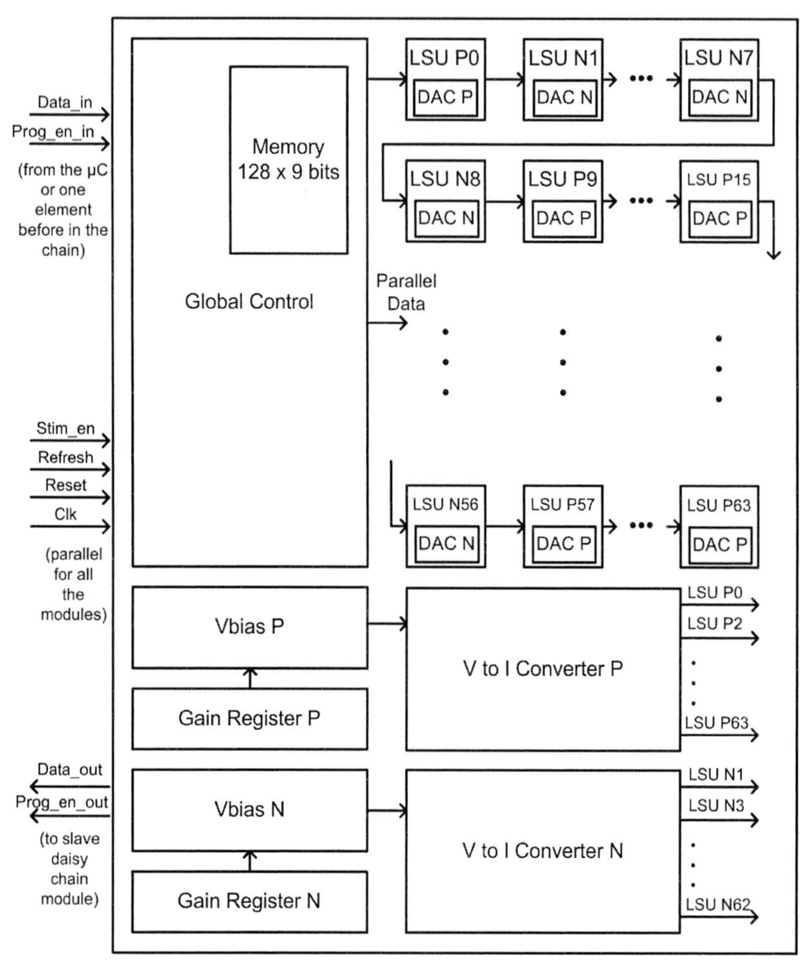

Figure 4.3.2 Block diagram, ASIC v3.

On chip there are two voltage bias modules, which bias the analog circuitry on the DACs, with programmable gain through digital registers. The voltage coming out from the modules will be converted into 64 bias current lines, one for each LSU.

LSU

The ASIC consists of a matrix of 64 LSUs: 32 P type and 32 N type, see Figure 4.3.2, adjacent LSUs are of different types. The connection registers define the LSU pairs, i.e. which LSU will interacts with each other, top, bottom, left or right. Each ASIC can control 64 electrodes with 32 stimulation sites.

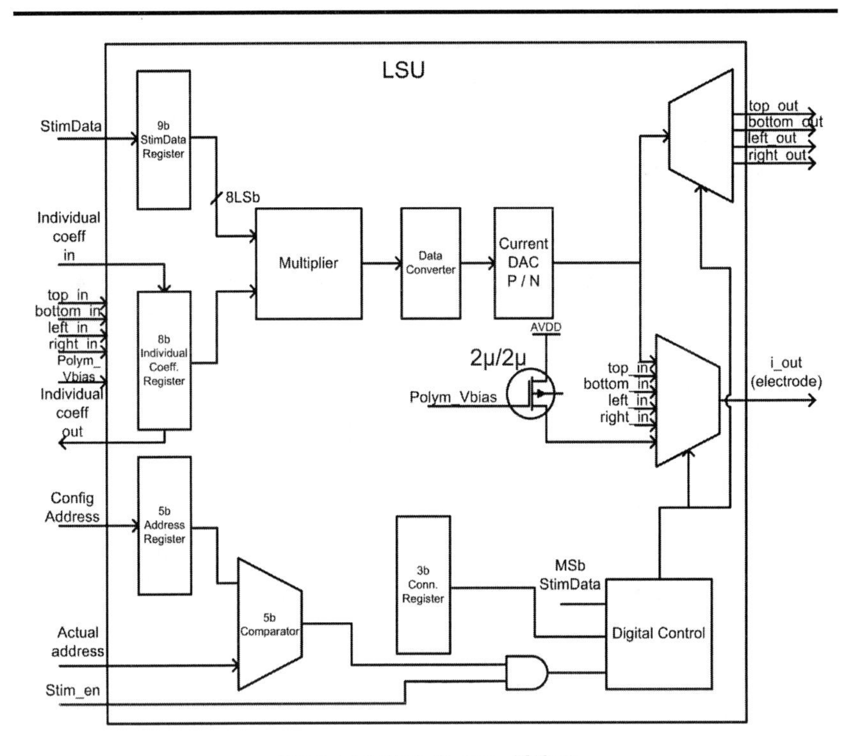

Figure 4.3.3 Block diagram, ASIC v3.

The LSU receives one 9 bit word, which is multiplied by the individual coefficient register (to give different amplitude to each LSU), the resulting data is converted into an analog signal by the DAC. The digital control inside of the LSU enables the output of the DAC, according the actual stimulating address and the configured in the register, this give the possibility to give to each LSU a different time slot. At the same time the digital control select the stimulating neighbor, according the configuration of the 3b connection register.

Digital Control

All the digital logic reacts to rising edge clock, and its inputs and outputs are active high. It has one clock input lines. The ASIC has three operation modes: Program, Polymerization and Stimulation.

Program mode

It receives 128 words (9 bits each, 1152 bits in total) for the memory (stimulation waveform), after that, 64 words (8 bits each, 5 bits stim address + 3 stim connection, 512 bits in total), then 25 bits for the polymerization counter, then 25 bits for the interpulse counter, 5 bits to determine the highest address, 4 bits for the N gain register, and 4 bits for the P gain register, In case of chips in daisy chain, the following data should to have the same format for the next chips.

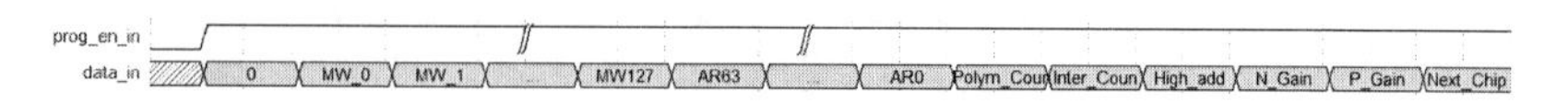

Figure 4.3.4 Program string, ASIC v3.

The 0 at the beginning is just a header bit. In all the words should come the LSB first.

Memory Word – 128 words, each one is composed of 9 bits, in sign magnitude format.

Address Register – 64 words, each one is composed of 8 bits. 3 MSB bits are the connection of the LSU, the 5 LSB are the address of the x LSU, it will be stored in the address register and indicates the time slot of the x LSU.

Table 4.3.1 Address Register configuration, ASIC v3.

AR[7:5]	Pair
000	Floating
001	Top
010	Bottom
011	Left
100	Right
101	VSS
110	Polymerization
111	Floating

Polymerization Counter – 25 Bits = number of clk cycles, it determines the duration of polymerization.

Interpulse counter – 25 bits = number of cycles that should wait before start the stimulation again.

Highest address – 5 bits, it determines how many time slots are given for the LSUs

N Gain and P Gain – 4 bits, to determine the N and P gain, respectively. 0000 – Lowest, 1000 – Middle, 1111 – Highest (0000 all switches are off, 0001 only the first switch is on).

Polymerization mode

To start the command the prog_en_in should go high and data_in should to be 1. It delivers current on each electrode output, all simultaneously or on different timeslots. At the beginning it puts all the individual coefficients to 0, then activate the polymerization (stim) for the corresponding time of the counter, for address 0, then 1, then 2, until the last address set by highest address register is reached, then ends. Before starting the polymerization command, the module should be programmed, with the address registers configured for the polymerization.

The time is determined by the polymerization counter register. The amplitude of the current will be determined by the voltage bias on the polymerization_bias pin, see Figure 4.3.5.

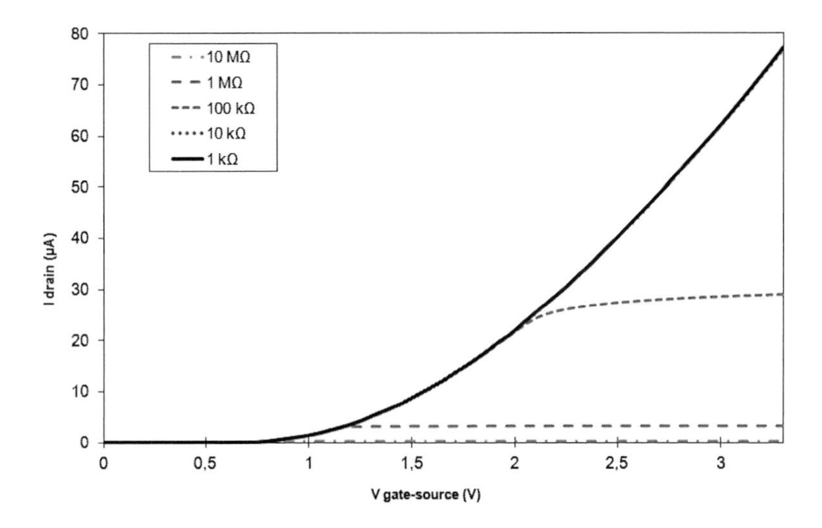

Figure 4.3.5 Transfer function of the polymerization transistor with different loads, ASIC v3.

From 100 kΩ loads, the transistor moves out of the saturation region. This is because with 3.3 V is not possible to satisfy the drain-source voltage requirement and the voltage drop at the load.

We need to remember the target is to polymerize metal electrodes. Since, we start with metal electrodes and then we will have PEDOT electrodes, the impedance will be changing during the process according to the interface between the electrode and the solution. To have a determined thickness of the PEDOT layer is necessary to perform a good galvanostatic deposition, which involves keeping the current constant.

Figure 2.2.14 showed the effect of the size in the PEDOT electrodes. From there, we can derivate the next two tables for micro and macro electrodes:

Table 4.3.2 Polymerization parameters of microelectrodes, ASIC v3.

Area	177 μm^2	413 μm^2	1000 μm^2
DC Gold Impedance	~10 MΩ	~7 MΩ	~5 MΩ
DC PEDOT Impedance	~1 MΩ	~400 kΩ	~200 kΩ
Current for polymerization (0.5 mA/cm^2)	0.88 nA	2.06 nA	5.00 nA
Necessary voltage at the gate of the polymerization transistor	493.23 mV	528.00 mV	564.54 mV

Table 4.3.3 Polymerization parameters of macroelectrodes, ASIC v3.

Area	0.5 mm²	1 mm²	2 mm²
DC Gold Impedance	~250 kΩ	~190 kΩ	~22 kΩ
DC PEDOT Impedance	~700 Ω	~430 Ω	~300 Ω
Current for polymerization (0.5 mA/cm²)	2.50 µA	5.00 µA	10.00 µA
Necessary voltage at the gate of the polymerization transistor	1.11 V	1.29 V	1.55 V

In the next two figures, it is possible to see a simulation of the polymerization transistor drain current under the necessary circumstances to polymerize micro and macro electrodes, respectively. The current will stay constant while the impedance is changing, because of the polymerization process.

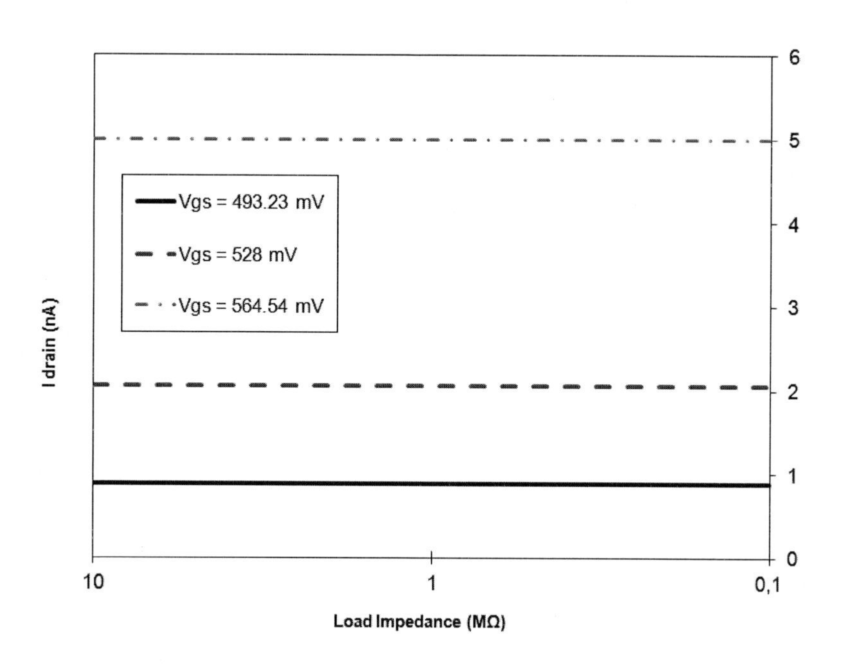

Figure 4.3.6 Polymerization transistor against different load impedances, for micro electrodes, ASIC v3.

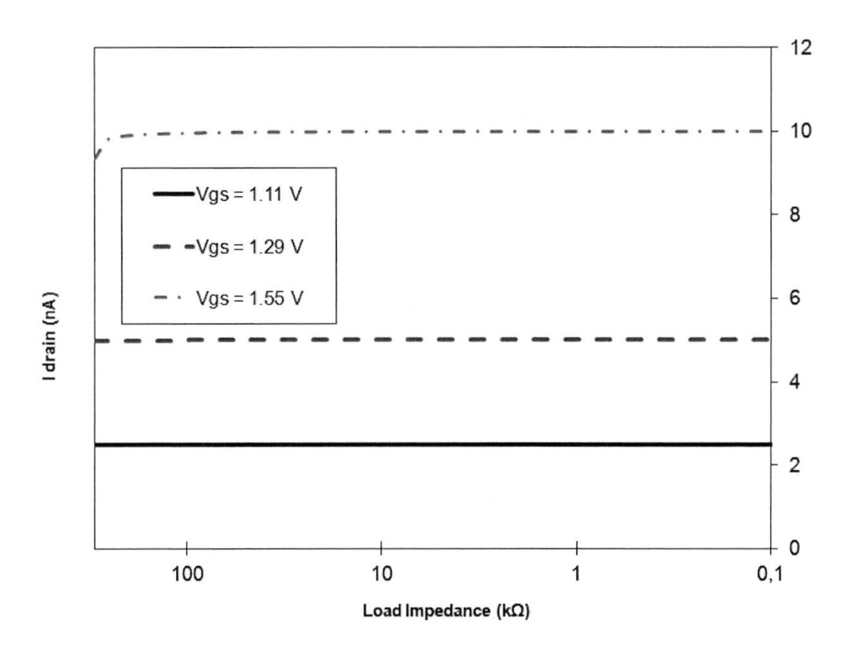

Figure 4.3.7 Polymerization transistor against different load impedances, for macro electrodes, ASIC v3.

Stimulation mode

To start this command is necessary to drive high the stim_en line. It starts the stimulation with the waveform stored in the memory. It begins by putting all the individual coefficients to 0, and then begins the cycle; it delivers the 128 words of the memory for the address (time slot) 0, it repeat for address 1 and so on until the highest address is reached. Then, it updates the individual coefficients (if a refresh command is present). Afterwards, it waits the time determined by the interpulse counter and starts the cycle again.

During the stimulation when the sign of the memory is positive (recommended for the first half of the stimulation waveform), the electrode is connected directly to the DAC on the LSU. When the sign of the memory is negative (recommended for the second half of the stimulation waveform) the electrode is connected to the neighbor cell determined by the connection register.

During the stimulation the individual coefficients on each LSU can be reprogrammed, i.e. change the stimulation amplitude of each stimulation site. Then it is possible to stimulate with a specific frequency and refresh the data with another frequency. This is useful, for example, in visual prostheses to create the images.

To reprogram the individual coefficients during the stimulation, it is necessary to drive the refresh line high and deliver a string of data. The string is received serially into the data_in (64 individual coefficients, 8 bits each = 512 bits), starting from LSU 63 and ending with LSU 0, the LSB should to be first. When the refresh line goes low again, the data will be refreshed on the registers. By giving a 0x00 to a LSU, it will be disabled. And with 0xFF, it will have the maximum amplitude.

Charge Balanced Stimulation

During the positive phase the electrodes are connected directly to their own DAC, and during the negative phase they will be connected through internal pass gates to their corresponding pair, see Figure 4.3.8. Each pair can have its own time slot. The charge imbalanced stimulation caused by process mismatch is avoided, because a pair of electrodes is driven always by the same current cells. And the same phenomenon caused by cross stimulation is prevented by stimulating on different time slots.

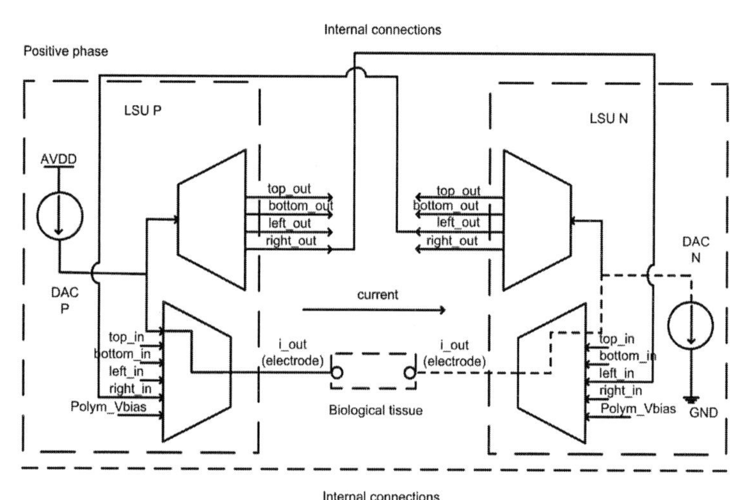

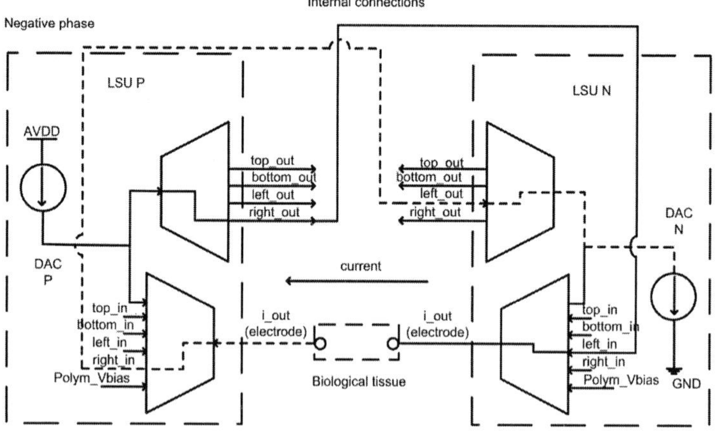

Figure 4.3.8 Example of positive and negative stimulation phase, ASIC v3.

DAC

The architecture implemented it is a hybrid current steering DAC, which performs the conversion directly from digital to current signal. The hybrid architecture is combining a binary weighted array and a unary array. This architecture allows having smaller DACs than the unary array and solves the glitch problems of the binary array. We have analyzed the architecture in the section 4.2.

Due to the equilibrium of number of cells we decided to use the combination of 3 unary bits and 5 binary bits. The DAC is formed by 12 current cells; 5 of them are binary weight scaled, where the smallest is equivalent to 1x LSB and the largest to 16x LSB; and 7 unary current cells each one represents 32x LSB. The number of bits of each DAC is 8, and because of the bipolar stimulation, the stimulator behaves like a 9 bit DAC. Figure 4.3.9 shows the structure of the P DAC's current cell. The N DAC is implemented by using the same architecture with N MOS transistors. In the N folded cascode equivalent to 1x LSB, the width of the transistors is 0.77 μm.

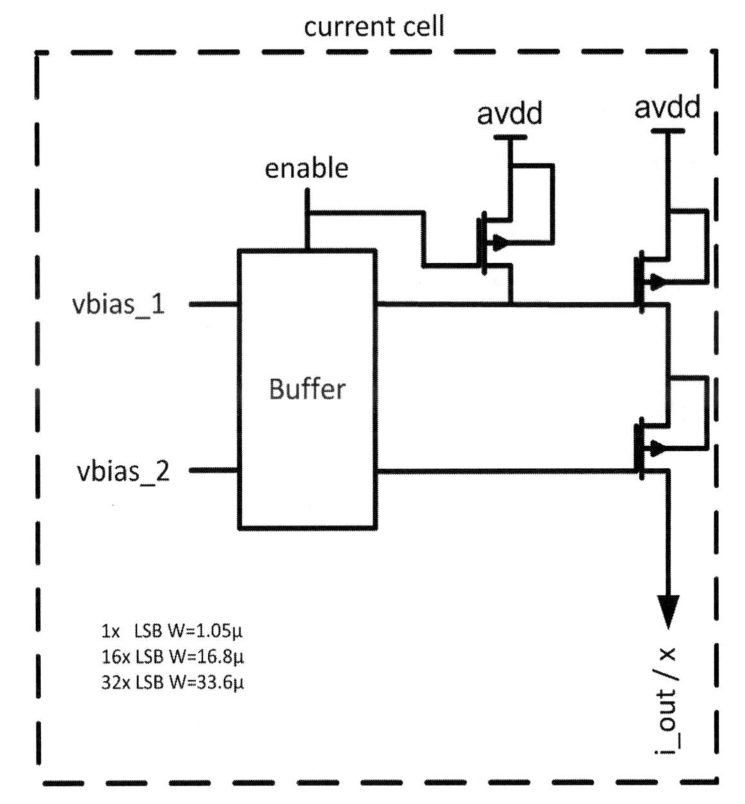

Figure 4.3.9 Current of the P DAC, ASIC v3.

Voltage Bias Circuit

In order to keep the bias voltage stable versus noise, changes of voltage from power supply, we implemented the same topology of the Beta-Multiplier circuit used in the ASIC version 2, Figure 4.2.6.

In this ASIC our design includes N and P DACs, they were matched in design. However, it is necessary to minimize the process mismatch. In order to have the ability to adjust individually each type of DAC we implemented two voltage bias circuits, for the P and for the N circuits. And we replaced the resistor that generates the reference current in the beta multiplier circuit, instead of it, we designed a variable resistor. We can change its total value by adding resistors in parallel to the main resistor. The schematic is shown in Figure 4.3.10.

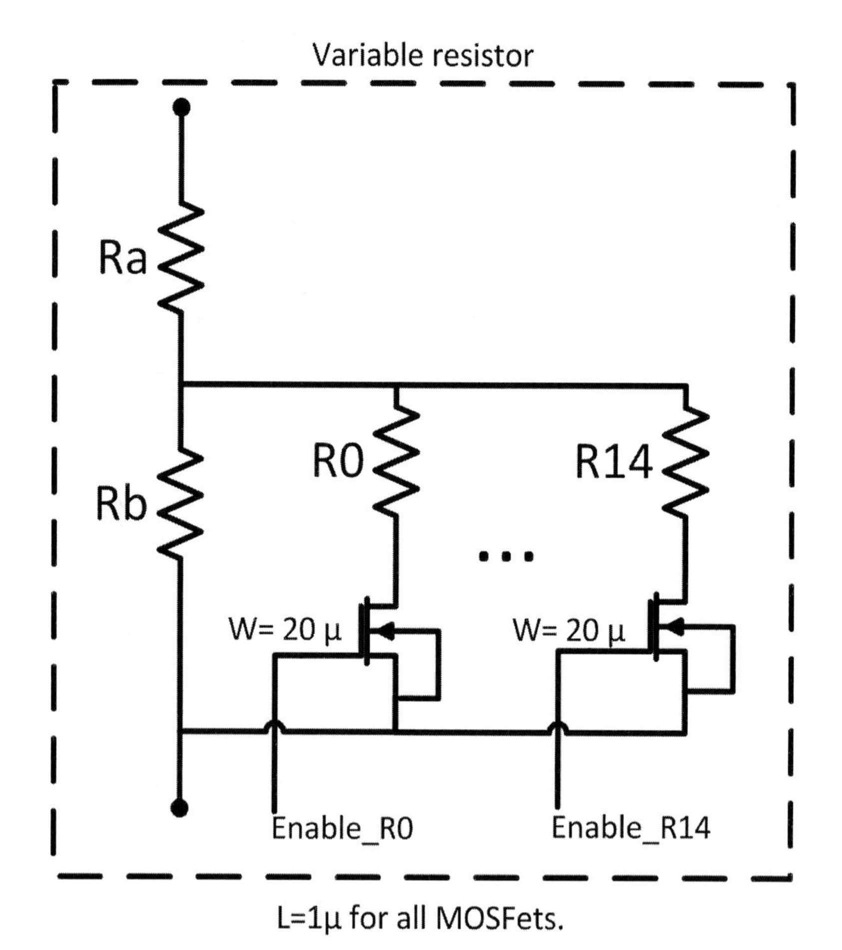

Figure 4.3.10 Diagram of the variable resistor on the voltage bias circuit, ASIC v3.

The total amount of resistors are 17, they are controlled digitally in thermometer code. We have a pure binary to thermometer code converter. Thus, we can control the bias circuit with a four bit register. The resulting ohmic values are depicted in Table 4.3.4.

Table 4.3.4 Control and resulting value of the variable resistor on the voltage bias circuit, ASIC v3.

Register Value (dec)	**Thermometer Code**	**Resistor**	Individual Resistor Value in P circuit	Resulting Value in P circuit	Individual Resistor Value in N circuit	Resulting Value in N circuit
		Ra	4.54 kΩ		4.54 kΩ	
0	000000000000000	Rb	960 Ω	5.50 kΩ	920 Ω	5.460 kΩ
1	000000000000001	R0	14.40 kΩ	5.44 kΩ	14.47 kΩ	5.405 kΩ
2	000000000000011	R1	12.60 kΩ	5.38 kΩ	12.74 kΩ	5.350 kΩ
3	000000000000111	R2	10.92 kΩ	5.32 kΩ	11.12 kΩ	5.295 kΩ
4	000000000001111	R3	9.36 kΩ	5.26 kΩ	9.61 kΩ	5.240 kΩ
5	000000000011111	R4	7.92 kΩ	5.20 kΩ	8.21 kΩ	5.185 kΩ
6	000000000111111	R5	6.60 kΩ	5.14 kΩ	6.92 kΩ	5.130 kΩ
7	000000001111111	R6	5.40 kΩ	5.08 kΩ	5.74 kΩ	5.075 kΩ
8	000000011111111	R7	4.32 kΩ	5.02 kΩ	4.67 kΩ	5.020 kΩ
9	000000111111111	R8	3.36 kΩ	4.96 kΩ	3.71 kΩ	4.965 kΩ
10	000001111111111	R9	2.52 kΩ	4.90 kΩ	2.86 kΩ	4.910 kΩ
11	000011111111111	R10	1.80 kΩ	4.84 kΩ	2.12 kΩ	4.855 kΩ
12	000111111111111	R11	1.20 kΩ	4.78 kΩ	1.49 kΩ	4.800 kΩ
13	001111111111111	R12	720 Ω	4.72 kΩ	970 Ω	4.745 kΩ
14	011111111111111	R13	360 Ω	4.66 kΩ	559 Ω	4.690 kΩ
15	111111111111111	R14	120 Ω	4.60 kΩ	259 Ω	4.635 kΩ

Figure 4.3.11 shows the output of the Vbias 1 against the different register values. And in Figure 4.3.12 is depicted the corresponding output of the Vbias 4.

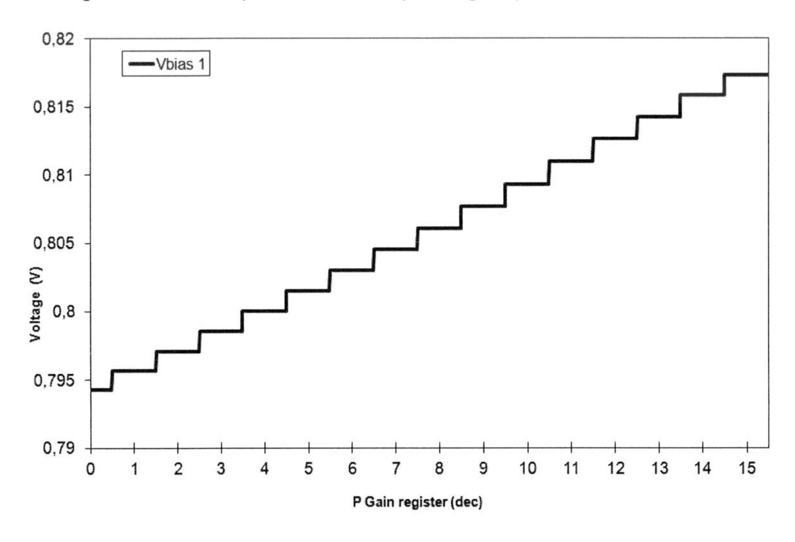

Figure 4.3.11 Vbias 1 against the different P Gain register configurations, ASIC v3.

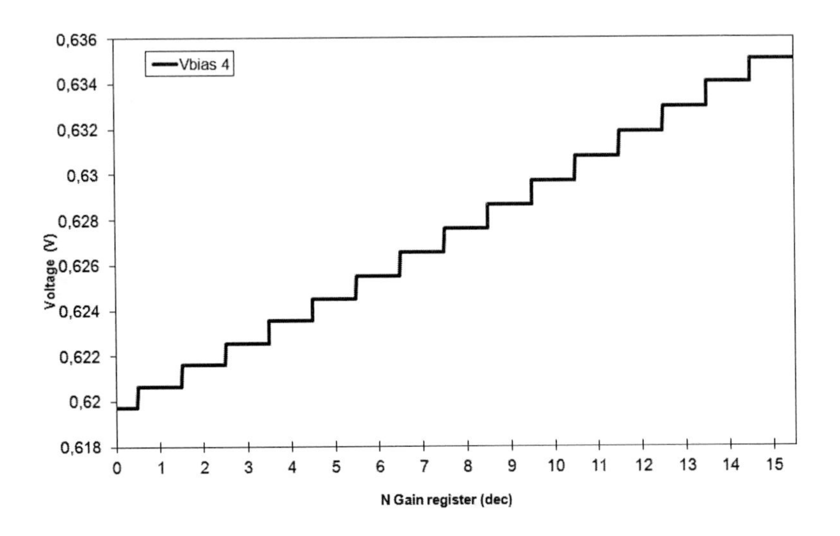

Figure 4.3.12 Vbias 4 against the different N Gain register configurations, ASIC v3.

Vbias 1 and Vbias 4 control the output of the P DAC and N DAC, respectively. Their normalized outputs under the different register configuration are shown in Figure 4.3.13 and Figure 4.3.14.

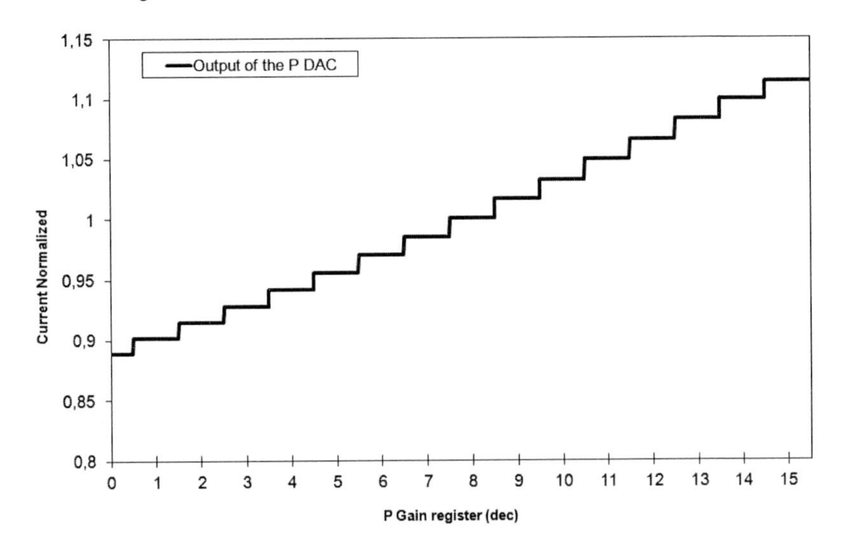

Figure 4.3.13 Normalized output of the P DAC against different P Gain register configurations, ASIC v3.

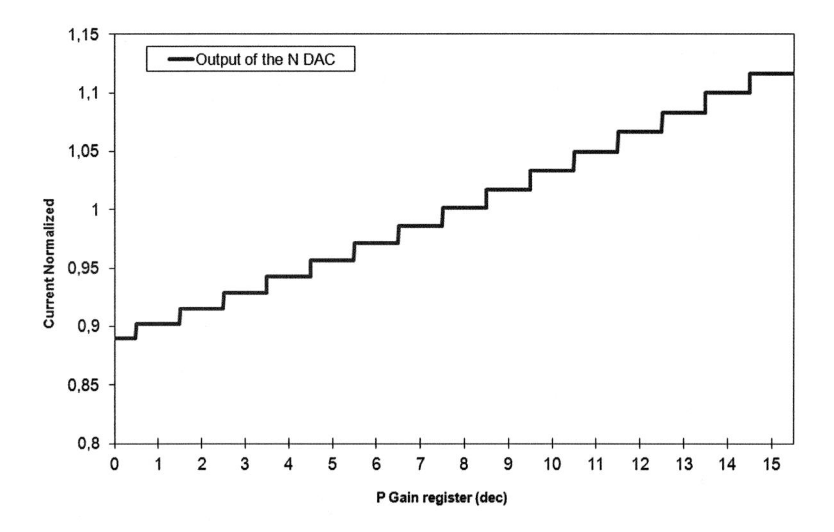

Figure 4.3.14 Normalized output of the N DAC against different N Gain register configurations, ASIC v3.

4.3.2. Experimental Results

Figure 4.3.15 shows a picture of the fabricated ASIC. The resulting total area is 4.25 mm² (1.924 µm x 2.209 µm); 0.039 mm² for each LSU (188 µm x 209 µm). For packaging information refer to appendix E.

An electrode array of 8 individual electrodes was polymerized by using the internal module of the ASIC. The array was 95 nm thick made of gold deposited over a 75 µm flexible polyimide thick substrate. The array was passivated to leave 1000 µm² area circular electrodes (35.7 µm diameter) with an electrode interdistance of 5 mm. To perform the PEDOT galvanostatic deposition the array was immersed in an EDOT/NaPSS micellar dispersion along with a gold counter electrode. The ASIC was programmed to deliver 5 nA for 360 s (180 mC/cm²) and the eight electrodes were covered of PEDOT one after other. The deposition of a blue material on the electrode surface confirmed the PEDOT deposition. The gold electrode fabrication and the PEDOT polymerization were performed according to the chapter 2.

The next configuration was used to perform the tests: 4 MHz ASIC clock, the stimulation waveform programmed was a symmetrical biphasic current pulse, anodic phase first, 288 µs width sinusoidal with a 2.5 ms interpulse without interphase. Three pairs of channels were activated with amplitudes of 50 µA, 20 µA and 30 µA, respectively. Each pair had its own time slot. For the measurements, the electrode array was immersed in a glass recipient containing saline solution (DeltaSelect GmbH), the impedance between two electrodes (including the solution resistance) was found to be around 27 kΩ at 1kHz.

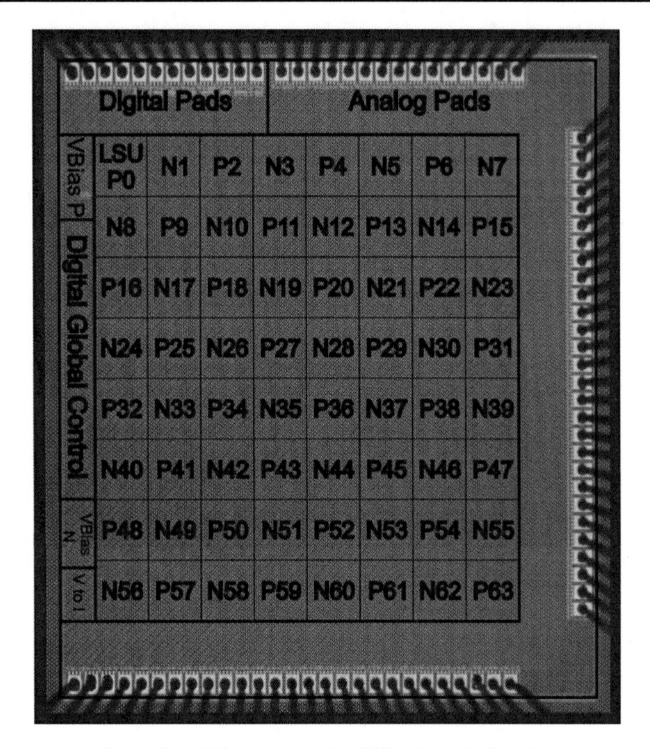

Figure 4.3.15 Micrograph of the ASIC v3 and its floorplan.

The power consumption of the stimulator was found to be 56.586 µW in standby mode, 3.451 mW in stimulation mode without load signal, 3.458 mW with load on the pair 1 and 3.468 mW with load on the three pairs of channels.

Figure 4.3.16 shows the current injected and the voltage measured at the electrodes, with the sinusoidal waveform described above and with a rectangular waveform.

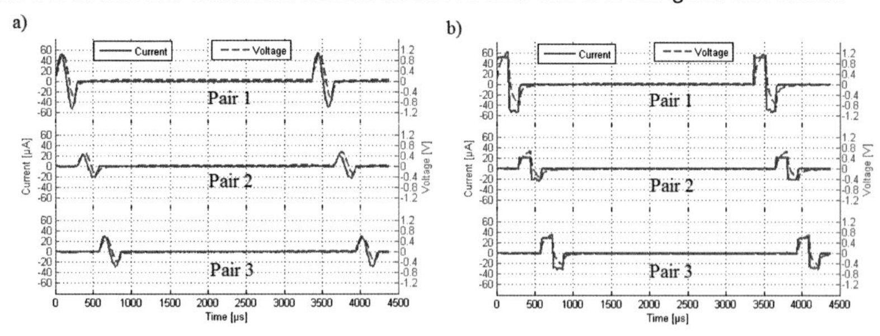

Figure 4.3.16 Current and voltage on PEDOT microelectrodes, ASIC v3.

By placing a variable resistor as load, the maximum output current was found to be ±55.0 µA and the maximum output voltage ±3.14 V.

Figure 4.3.17 shows the charge accumulated on time for a single pulse. The residual charge after one second of stimulation with the sinusoidal waveform for the pair 1 was found to be around +4.70 nC. Due to the offset on the acquisition card, the measurement was done as follows: with the stimulator unconnected, the current flowing through the electrodes was recorded for one second, and the residual charge was calculated as offset; then, with the stimulator attached to the three pairs, the current flowing through the electrode pair 1 was recorded for one second, and the residual charge was calculated as total charge; afterwards the offset was subtracted from the total charge, the same procedure was repeated 50 times, and the results were averaged.

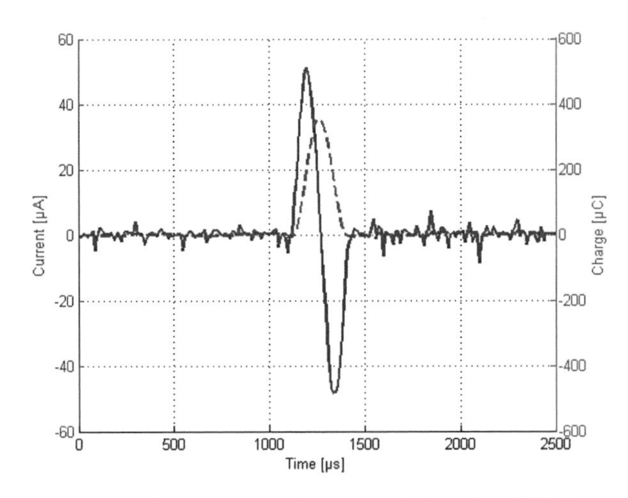

Figure 4.3.17 Residual charge after one stimulation pulse, ASIC v3.

Table 4.3.5 summarizes results and specifications of the ASIC.

Table 4.3.5 Comparison of ASIC v3 specifications.

	[117] **2012**	[118] **2011**	**Hybrid Architecture**
Technology	350 nm CMOS	65 nm CMOS	130 nm CMOS
ASIC area	--	25 mm^2	4.25 mm^2
Stimulator area	--	3.33 mm^2	2.84 mm^2
Supply voltage	3.3 / 20 V	1.0 / 3.3 V	1.2 / 3.3 V
Number of channels	2	64	64
Scalable	No	No	Yes, daisy chain
Power consumption	--	--	3.45 mW [a]
		LSU	
Area	0.235 mm^2	0.040 mm^2	0.039 mm^2
Power consumption per LSU	330 μW [a,b]	200 μW [a]	54 μW [a]
Architecture	DAC, delivering different waveforms	Current source, deliver only rectangular waveforms	Hybrid current steering DAC, delivering different waveforms
Resolution	5 Bits	64 Amplitude levels	8 / 9 Bits
Max. output	±10 V / programmable ±60, 120, 240, 480 μA	±3.3 V / --	±3.14 V / ±55.0 μA
LSB	programmable 4, 8, 16, 32 μA	--	210.00 nA

a. Without load, b. Simulated with 5 V as VDD.

4.3.3. Discussion

By using our design, and attaching 16 chips in daisy chain configuration, the total area requirement will be 68 mm^2. The power consumption will be around 55.2 mW. This arrangement is able to drive up to 1024 electrodes independently. This configuration gives flexibility, on one hand, it makes possible to locate each chip on different position, so close to the stimulation group as possible, and all the stimulation sites will be perfectly synchronized. On the other hand it is not necessary to have a big single silicon piece, it is possible to attach the multiple chips in a 3d chip stacking, thus, save area by giving to the system more volume.

In [116] it is shown that residual dc current higher 100 nA is correlated with neural tissue damage. The net imbalance is because current source and sink drivers are typically not perfect matched and the flow of current through cross electrodes is present. Thus, by exchanging the P and N cells between the pair of stimulation electrodes, and by assigning different time slots to each electrode pair, the residual charge decreases. The design presents a residual charge within the safe levels, 4.70 nC per second, i.e. 4.7 nA.

4.4. Chapter Conclusion

In this chapter, it was seen the steps of the design of a neurostimulator. Starting from the basis of the analog output on ASIC version 1, from where the following concepts were rescued.

The current steering architecture is ideal for the application because of the necessity for current (rather than voltage) at the output, its low power consumption, its capability of sharing some common state for a multiarray stimulator, and also for its simplicity.

The use of a cascode for the current cells improves its output resistance, presents a high voltage dynamic range and presents also low power consumption because the current is flowing through the branch only when the current cell is turned on.

The switches that interrupt the current's flow are located normally between the load and the current cell, we chose to put them at the transistor gates in order to minimize the number of transistors in the branch to avoid voltage drops.

In the ASIC 2 we presented a chip with 4 stimulator channels. The focus was the integration of the system. It was reached through the hybrid architecture, which retains the benefits of the unary current steering architecture in combination with smaller chip area. The output voltage was increased by migrating the design to thick oxide transistors. The stability of the implemented voltage bias circuit versus supply voltage changes was demonstrated, as well as the versatility of working with different supply voltages in order to reduce the power consumption according to the output voltage requirements.

The ASIC version 3 showed a fully implantable chip suitable for invasive neurostimulation. The design is scalable by attaching multiple chips in daisy chain configuration, and the six control lines remain the same for one or several chips. The stimulator shows flexibility regarding the stimulation parameters and the waveform. During the stimulation it is possible to synchronically refresh the stimulation amplitude on each individual LSU, even on every chip in daisy chain configuration, which is especially useful for applications such as retina implants.

The presented schema helps to maintain the residual charge within safe levels.

An on-chip module was implemented, capable of controlling the electropolymerization process of the electrodes.

The use of small transistor technology and the hybrid architecture of the DAC makes possible the integration of high number of LSUs in a single chip and the low power consumption.

5. Conclusion

This dissertation has shown the design of an improved neural stimulator through the use of several disciplines and techniques.

Regarding the electrodes, PEDOT shows good electrochemical properties, lower impedance than uncoated electrodes, better charge injection capability, the ability to improve the cell interaction by attaching some agents, and good stability. PEDOT is therefore an attractive candidate for invasive stimulation microelectrodes.

PEDOT could be deposited using different methods, such as spin coating and electrochemical polymerization. However, because of its morphology, its electrical anisotropic conductivity and the necessity for special compatible lithographic solvents, the PEDOT films deposited by spin coating seem to be inadequate for neural stimulation electrodes. Thus, PEDOT can be polymerized directly on the electrodes by electrochemical polymerization.

The electrochemical deposition is a simple method to deposit the PEDOT layer on the electrodes, and it is possible even when the implants are already assembled so long as the stimulator circuitry is able to deliver the current under the necessary parameters to perform the electropolymerization.

Regarding the polymerization parameters, low current densities are recommended to avoid substrate degradation and to improve the polymer package. The transferred charge is found to be proportional to the thickness of the resulting layer. Thicker layers improve the Q_{inj}, and reduce the impedance in the interface. However, it is necessary to find the optimal point where Q_{inj} and impedance are improved and the layer does not show delamination.

PEDOT/NaPSS electrical properties (Impedance and Q_{inj}) were constant and reproducible after several experiments, even *In Vivo*. These parameters can be estimated for macro and micro electrodes in function of the charge density.

It was also shown that by using a passive membrane model, energy can be saved by performing stimulation with non-rectangular waveforms. However, in order to have more accurate results, a second model was developed to incorporate equivalent circuits for the electrodes and tissues along the path.

The second model is an equivalent circuit of the elements in the path from the electrode to the cell membrane. This equivalent circuit model was divided by tissues, the aim of which is to facilitate future research analysis of biological models. The removal of sub-equivalent circuit models can be done and can simulate nerve stimulation from another specific point, e.g. below the stratum corneum.

A setup for experiments was developed, i.e. portable neuromuscular stimulator and its graphical user interface. Its flexibility makes the device suitable for performing experiments with changing parameters. The system enables experimentation not only with the four mentioned waveforms, but also allows experimentation with continuous stimulation pulses and the comparison of different waveforms against the changing of parameters such as interphase, distance between pulses or width of the train. The size and weight of the stimulator make it suitable for daily use as FES. And its design also serves as the basis for the further development of a multielectrode stimulator.

By *In Vivo* experimentation, the results of single pulse stimulation show more efficient neuromuscular stimulation by using different waveforms rather than the rectangular pulse, which is nowadays the most widely used waveform. The tendency of the curves is consistent over several days, using multiple test subjects. The required energy is lower in all cases. The voltage is also lower in most cases. Each waveform exhibits its own attributes. The waveform could be selected according to the specifications of the system and according to the requirements for either lower currents or lower voltages.

The optimal pulse width was found to be around 256 μs, according to the current, voltage, charge injection and energy requirements.

The required charge injection at the electrodes was lower by using non-rectangular waveforms. By knowing the architecture of the stimulator output stage, the charge injection requirement is a metric that permits the calculation of energy required by the whole system.

Regarding the circuitry of the stimulator, the current steering architecture is ideal for the application because of the necessity for current (rather than voltage) at the output, its low power consumption, its capability of sharing some common state for a multiarray stimulator, and also for its simplicity.

The hybrid architecture retains the benefits of the unary current steering architecture in combination with smaller chip area.

The stability of the implemented voltage bias circuit versus supply voltage changes was demonstrated, as well as the versatility of working with different supply voltages in order to reduce the power consumption according to the output voltage requirements.

The ASIC version 3 showed a fully implantable chip suitable for invasive neurostimulation. The design is scalable by attaching multiple chips in daisy chain configuration, and the six control lines remain the same for one or several chips. The stimulator shows flexibility regarding the stimulation parameters and the waveform. During the stimulation it is possible to synchronically refresh the stimulation amplitude on each individual LSU, even on every chip in daisy chain configuration, which is especially useful for applications such as retina implants.

The presented schema helps to maintain the residual charge within safe levels.

An on-chip module was implemented, capable of controlling the electropolymerization process of the electrodes.

The use of small transistor technology and the hybrid architecture of the DAC makes possible the integration of a high number of LSUs in a single chip and low power consumption.

Appendix

A. *Microelectrodes on rigid substrate*

Electrodes for polymerization experiments were developed at the facilities of the university. The contact area of the electrodes was a gold layer over a silicon substrate. The electrodes were fabricated according the following figure:

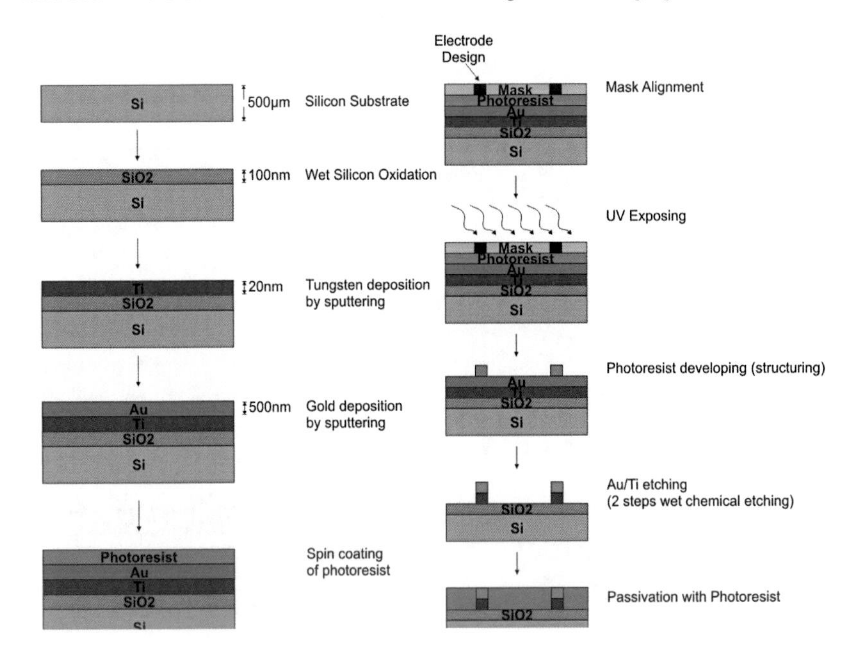

Figure A.1 Manufacturing process of microelectrodes.

The design of the electrodes is shown in the next figures, (all the distances are in micrometers). A passivation layer was applied on the structures in order to isolate them, the gold surface of the electrodes and the contacts were remaining free of isolation.

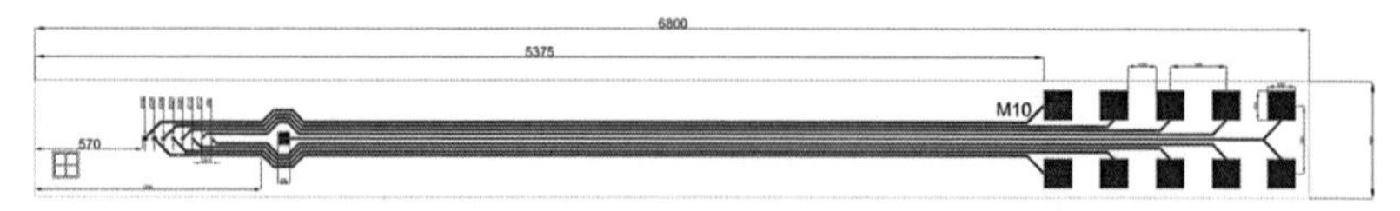

Silicon substrate with gold electrodes and contacts

Figure A.2 Silicon substrate with gold electrodes and contacts.

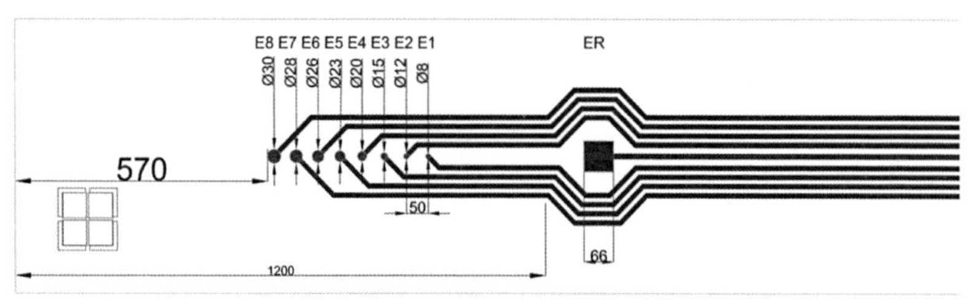

Electrodes surface

Figure A.3 Microelectrodes surface.

A PCB for mounting the silicon substrate and perform the tests was also designed and fabricated. The silicon substrate was glued on the PCB, and then was contacted through wire bonding. The contacts were covered with epoxy to isolate them.

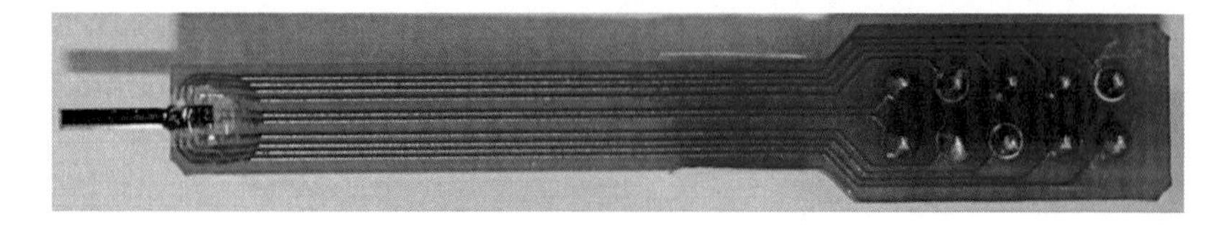

Figure A.4 Microelectrodes substrate mounted on a PCB.

B. Electrodes on flexible substrate

B.1. Electrodes for *In Vivo* Experimentation

A design of an array of flexible electrodes was developed for *In Vivo* experimentation. All the fabrication processes were performed in the facilities of the university.

The fabrication of gold electrodes was performed in an ultra-high vacuum chamber. A polyimide film 25 μm thick was used as substrate. In order to improve adhesion, a 5 nm chromium layer was deposited on the substrate.

The gold layer deposition was structured with a shadow mask. The substrate and the shadow mask were introduced in the chamber. The 95 nm thick gold layer was deposited by electron beam evaporation of a high purity metal in a background of 6.66×10^{-4} Pa. The deposition rate was on the order of 3 Å/s.

Following deposition, a positive photo-resist (S-1805) layer was applied. A film mask was aligned with the design, and the samples were exposed to UV light. Then, the photo-resist was developed, leaving non-insolated only the desired area for the electrodes (1 mm^2) and the pads exposed for contacting them.

A small PCB was developed, it had the contact pads. In was put on the substrate pads with soldering paste in between, and then annealed until the paste was melted. Then the PCB was covered with glue to isolate the connection.

On the array, some of the electrodes were covered with PEDOT, polymerization parameters: current density of 0.5 mA/cm^2 and charge of 60 mC/cm^2. And some electrodes remained as gold electrodes.

Figure B.1 Electrodes on flexible substrate for *In Vivo* experimentation.

B.2. Electrodes for Experimentation

A second version of the flexible electrodes was developed. The design of the array was modified by placing electrodes with larger areas of metal (2 mm²) and changing the method of the contact. In the last design, the metal of the electrode area was fixed and the exposed area was a little smaller, in the new design the metal area was increased in order to use several photomasks with different exposed areas (177 µm², 413 µm², 1000µm², 0.5 mm², 1.0 mm² and 2.0 mm²). Regarding the contact method, in the last design the array was soldered and glued, in the new design it was possible to contact it through a connector for its easy and fast change. This second version of the flexible electrode was useful for performing experimentation with different parameters of polymerization and electrode sizes. The fabrication process was the same.

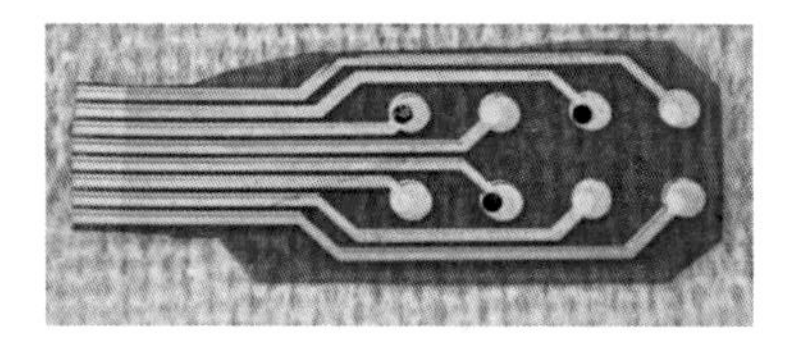

Figure B.2 Second version of flexible electrodes for experimentation.

C. *ASIC Version 1*

The die was packaged in a JLCC44, the bond diagram is shown in Figure C.1. The pin list is described in Table C.1.

.

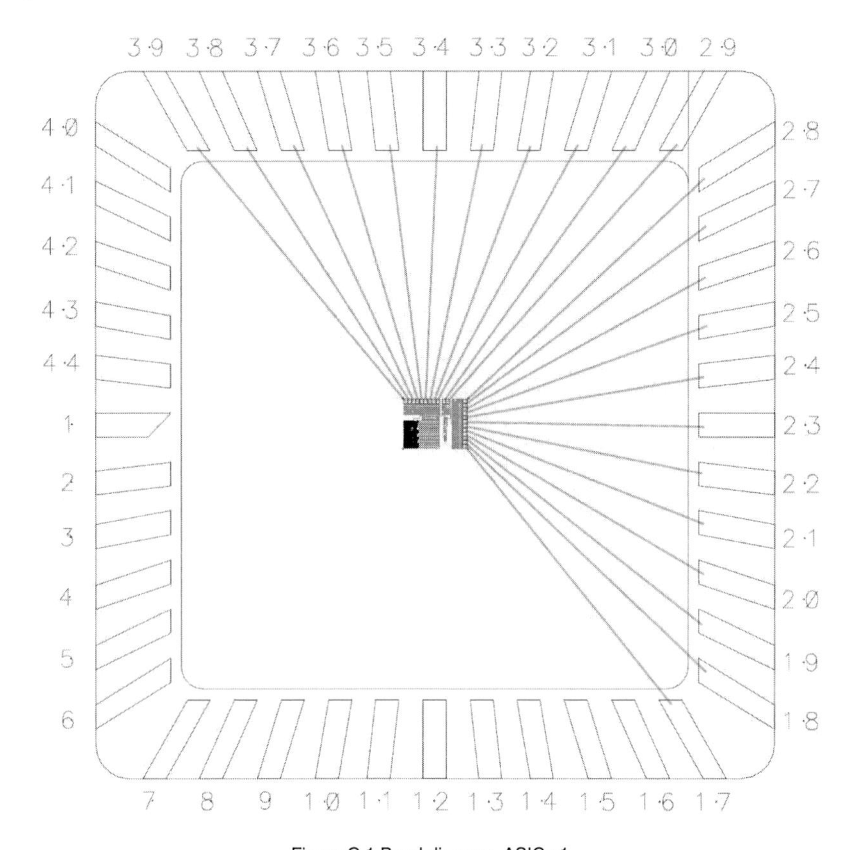

Figure C.1 Bond diagram, ASIC v1.

Table C.1 Pin list, ASIC v1.

Pin	Name	Type	Direction	IO_Cell	Function
1		NC			
2		NC			
3		NC			
4		NC			
5		NC			
6		NC			
7		NC			
8		NC			
9		NC			
10		NC			
11		NC			
12		NC			
13		NC			
14		NC			
15		NC			
16		NC			
17		NC			
18	e2	Analog	Inout		Electrode 2
19	e1	Analog	Inout		Electrode 1
20	IBPOST_4u	Analog	In		Current bias for analog channel
21	Vout_CH1	Analog	Out		Vout of analog channel
22	0V6	Analog	In		Voltage bias for analog channel
23	Vp_CH1	Analog	In		Positive input of analog channel
24	Vn_CH1	Analog	In		Negative input of analog channel
25	ILP_250n	Analog	In		Current bias for analog channel
26	IPRE_5u	Analog	In		Current bias for analog channel
27	VSS_ANA	Power	Inout		Analog ground for stimulator, 1.2 V
28	VDD_ANA	Power	Inout		Analog power for stimulator, 1.2 V
29	VSS_DIG	Power	Inout		Digital ground for stimulator, 1.2 V
30	VDD_DIG	Power	Inout		Digital power for stimulator, 1.2 V
31	p_start	Digital	In		Control line for starting the stimulation
32	p_reset	Digital	In		Global reset for the stimulator
33	p_data_in	Digital	In		Serial input data line for the stimulator
34	p_clk	Digital	In		Clock for the stimulator
35	p_busy	Digital	Out		State indicator of the stimulator
36	VSS_IO_DIG	Power	Inout		Digital ground of the IOs, 3.3 V
37	VDD_IO_DIG	Power	Inout		Digital power of the IOs, 3.3 V
38	VSS_CH	Power	Inout		Ground of the analog channel, 3.3 V
39	VDD_CH	Power	Inout		Power of the analog channel, 3.3 V
40		NC			
41		NC			
42		NC			
43		NC			
44		NC			

D. ASIC Version 2

The ASIC version 2 has four stimulation channels, a test channel for performing tests on the individual stages, called TS, and some test transistors. The diagram of the whole system is depicted in Figure D.1.

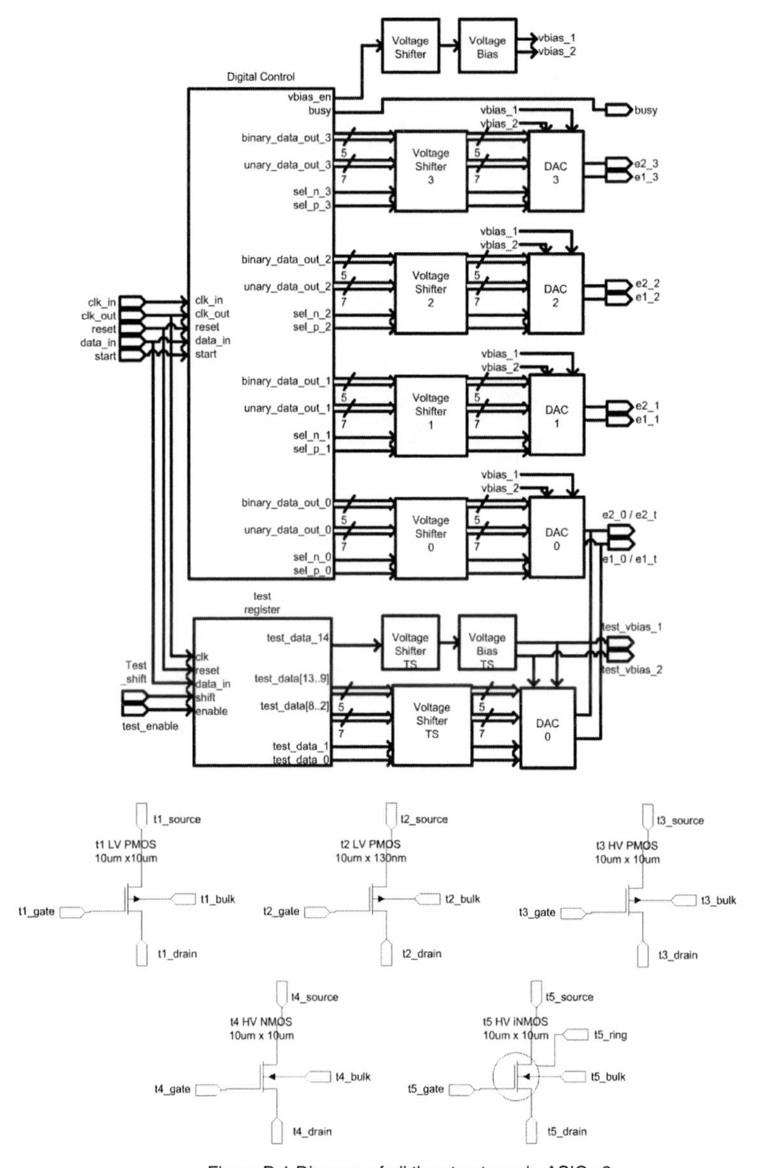

Figure D.1 Diagram of all the structures in ASIC v2.

The die was packaged in a JLCC44, the bond diagram is shown in Figure D.2. The pin list is described in Table D.1.

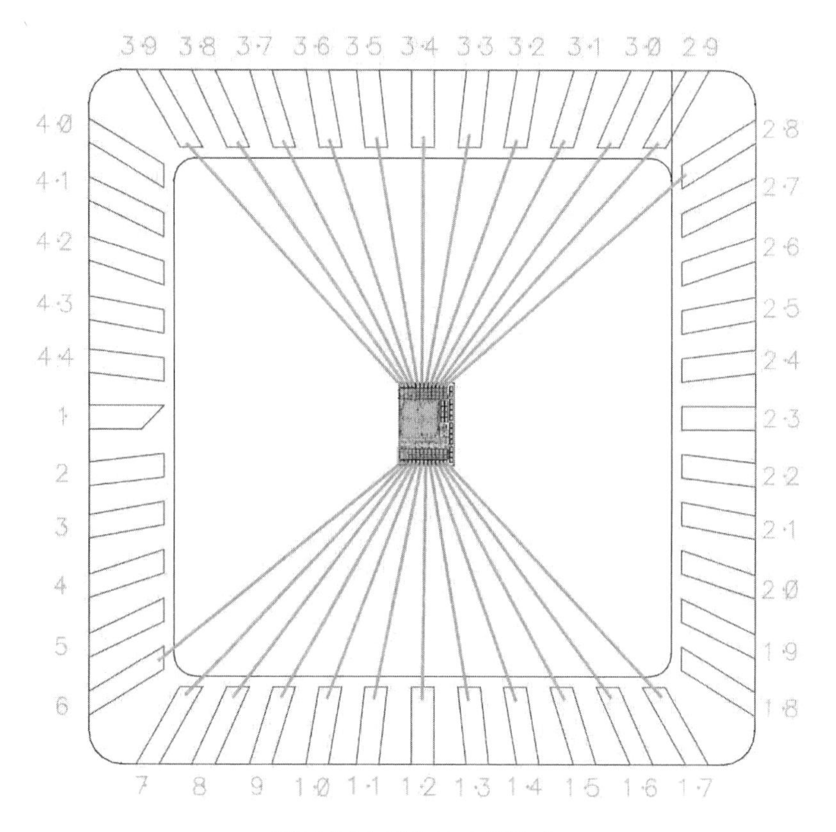

Figure D.2 Bond diagram, ASIC v2.

Table D.1 Pin list, ASIC v2.

Pin	Name	Type	Direction	IO_Cell	Function
1		NC			
2		NC			
3		NC			
4		NC			
5		NC			
6	avdd	Power	Inout		3.3 V Analog Power Supply
7	avss	Power	Inout		Analog Ground
8	e2_3	Analog	Inout		Electrode 2, channel 3
9	e1_3	Analog	Inout		Electrode 1, channel 3
10	e2_2	Analog	Inout		Electrode 2, channel 2
11	e1_2	Analog	Inout		Electrode 1, channel 2
12	e2_1	Analog	Inout		Electrode 2, channel 1
13	e1_1	Analog	Inout		Electrode 1, channel 1
14	e2_0 / e2_t	Analog	Inout		Electrode 2, channel 0 / test channel
15	e1_0 / e1_t	Analog	Inout		Electrode 1, channel 0 7 test channel
16	test_vbias_2	Analog	Out		Voltage sense for vbias_2 of test structure
17	test_vbias_1	Analog	Out		Voltage sense for vbias_1 of test structure
18		NC			
19		NC			
20		NC			
21		NC			
22		NC			
23		NC			
24		NC			
25		NC			
26		NC			
27		NC			
28	test_shift	Digital	In		Test register shift function
29	test_enable	Digital	In		Test register write enable
30	clk_in	Digital	In		Clock input for the programming stage
31	clk_out	Digital	In		Clock input for the stimulation stage
32	reset	Digital	In		Global reset
33	data_in	Digital	In		Serial input data line
34	busy	Digital	Out		State indicator
35	start	Digital	In		Control line for starting the stimulation
36	dvss	Analog	Inout		Digital Ground
37	dvdd	Analog	Inout		1.2 V Digital Power Supply
38	vssio	Analog	Inout		Ground for the IO-Pads
39	vddio	Analog	Inout		3.3 V Supply for the IO-Pads
40		NC			
41		NC			
42		NC			
43		NC			
44		NC			

The Test Structures have the purpose of testing each stage in individual form and consist of:

- Test register, which store the configuration for the test structures.
- Voltage Shifter, the same function as in SV2.
- Voltage Bias Circuit, the same function as in SV2, and it has also their output connected to external pins in order to test the module.
- One 8 bit DAC, it is the same design of SV2 and shares the output pins with one channel of SV2, thus, it must be careful with this DAC, it should not be activated both of them DACs.

Test register, it is a 15 bit double register. When the *test_shift* line is active, the first register receives the information in serial way, the LSB should be sent first. And when the line *test_enable* is active, the output is updated by transferring in parallel the data to the second register. The outputs are connected to the lines of the test structures according the following table:

Table D.2 Test register configuration, ASIC v2.

Bit	Description	Bit	Description	Bit	Description	Bit	Description
		11	DAC_binary_data_2	7	DAC_unary_data_1	3	DAC_unary_data_5
14	vbias_en	10	DAC_binary_data_3	6	DAC_unary_data_2	2	DAC_unary_data_6
13	DAC_binary_data_0	9	DAC_binary_data_4	5	DAC_unary_data_3	1	hbridge_sel_n
12	DAC_binary_data_1	8	DAC_unary_data_0	4	DAC_unary_data_4	0	hbridge_sel_p

The bits 2 to 13 are the digital input for the DAC in the test structure. The binary data corresponds to the LSB in the DAC, they should be feed in binary pure code. The unary data corresponds to the MSB, they should be feed in 7 bits thermometer code. The function of the hbridge follows the table:

Table D.3 H-Bridge function, test structures on ASIC v2.

hbridge_sel_n	hbridge_sel_p	Function
0	0	Pins disconnected
0	1	Positive stimulation (from e1 to e2)
1	0	Negative stimulation (from e2 to e1)
1	1	FORBIDDEN

It is important to notice that is completely FORBIDDEN to activate simultaneously the hbridge_sel_n and the hbridge_sel_p lines, this produces a short circuit between *avdd* and *avss.* Since the output pins (number 14 and 15) of the TS DAC and the DAC of the channel 0 in SV2 are the same, it is also FORBIDDEN the use of both DACs simultaneously.

The Test Transistors enable to perform characterization of these devices, the elements contained are:

- (t1)Low Voltage PMOS 10 µm x 10 µm.
- (t2)Low Voltage PMOS 10 µm x 130 nm.
- (t3)High Voltage PMOS 10 µm x 10 µm.
- (t4)High Voltage NMOS 10 µm x 10 µm.
- (t5)High Voltage isolated NMOS 10 µm x 10 µm.

They are not bonded to the package. *t4_source* is internally connected to *avss.* And the pads are:

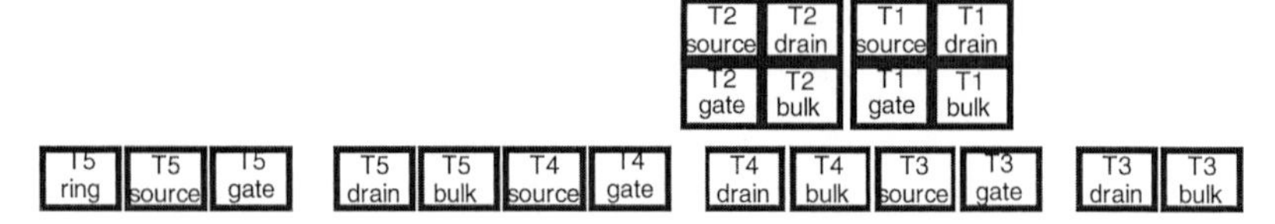

Figure D.3 Test transistors pads, ASIC v2.

E. ASIC Version 3

The die was packaged in a QFP120, the bond diagram is shown in Figure E.1. The pin list is described in Table E.1.

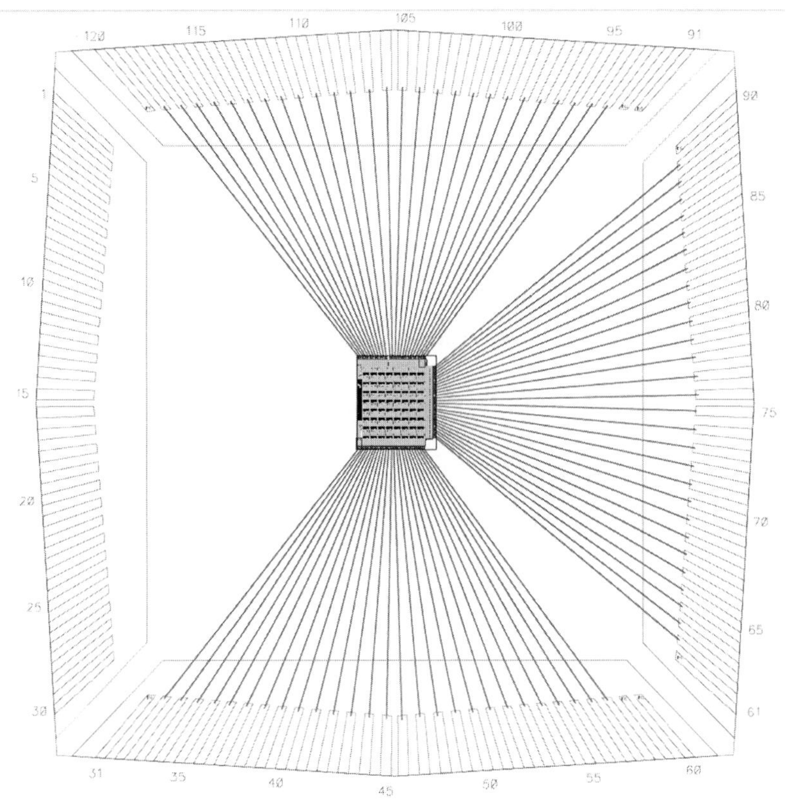

Figure E.1 Bond diagram, ASIC v3.

Table E.1 Pin list, ASIC v3.

Pin	Name	Type	Direction	IO_Cell	Function
1		NC			
2		NC			
3		NC			
4		NC			
5		NC			
6		NC			
7		NC			
8		NC			
9		NC			
10		NC			
11		NC			
12		NC			
13		NC			
14		NC			
15		NC			

16		NC			
17		NC			
18		NC			
19		NC			
20		NC			
21		NC			
22		NC			
23		NC			
24		NC			
25		NC			
26		NC			
27		NC			
28		NC			
29		NC			
30		NC			
31		NC			
32	avdd	Power	Inout		3.3 V Analog Power Supply
33	avss	Power	Inout		Analog Ground
34	i61	Analog	Inout		LSU Output
35	i48	Analog	Inout		LSU Output
36	i40	Analog	Inout		LSU Output
37	i32	Analog	Inout		LSU Output
38	i24	Analog	Inout		LSU Output
39	i16	Analog	Inout		LSU Output
40	i8	Analog	Inout		LSU Output
41	i0	Analog	Inout		LSU Output
42	i56	Analog	Inout		LSU Output
43	i49	Analog	Inout		LSU Output
44	i41	Analog	Inout		LSU Output
45	i33	Analog	Inout		LSU Output
46	i25	Analog	Inout		LSU Output
47	i17	Analog	Inout		LSU Output
48	i9	Analog	Inout		LSU Output
49	i1	Analog	Inout		LSU Output
50	i57	Analog	Inout		LSU Output
51	i50	Analog	Inout		LSU Output
52	i42	Analog	Inout		LSU Output
53	i34	Analog	Inout		LSU Output
54	i26	Analog	Inout		LSU Output
55	i18	Analog	Inout		LSU Output
56	i10	Analog	Inout		LSU Output
57	i2	Analog	Inout		LSU Output
58	i58	Analog	Inout		LSU Output
59		NC			
60		NC			
61		NC			
62	i60	Analog	Inout		LSU Output
63	i63	Analog	Inout		LSU Output
64	i62	Analog	Inout		LSU Output
65	i52	Analog	Inout		LSU Output
66	i55	Analog	Inout		LSU Output
67	i54	Analog	Inout		LSU Output
68	i53	Analog	Inout		LSU Output
69	i44	Analog	Inout		LSU Output
70	i47	Analog	Inout		LSU Output
71	i46	Analog	Inout		LSU Output
72	i45	Analog	Inout		LSU Output
73	i39	Analog	Inout		LSU Output
74	i38	Analog	Inout		LSU Output

75	i37	Analog	Inout		LSU Output
76	i31	Analog	Inout		LSU Output
77	i30	Analog	Inout		LSU Output
78	i29	Analog	Inout		LSU Output
79	i23	Analog	Inout		LSU Output
80	i22	Analog	Inout		LSU Output
81	i21	Analog	Inout		LSU Output
82	i15	Analog	Inout		LSU Output
83	i14	Analog	Inout		LSU Output
84	i13	Analog	Inout		LSU Output
85	i7	Analog	Inout		LSU Output
86	i6	Analog	Inout		LSU Output
87	i5	Analog	Inout		LSU Output
88	polymerizatio n_bias	Analog	In		Bias Voltage for the polymerization module
89	avss	Power	Inout		Analog Ground
90		NC			
91		NC			
92		NC			
93	avss	Power	Inout		Analog Ground
94	avdd	Power	Inout		3.3 V Analog Power Supply
95	i4	Analog	Inout		LSU Output
96	i12	Analog	Inout		LSU Output
97	i20	Analog	Inout		LSU Output
98	i28	Analog	Inout		LSU Output
99	i36	Analog	Inout		LSU Output
100	i59	Analog	Inout		LSU Output
101	i3	Analog	Inout		LSU Output
102	i11	Analog	Inout		LSU Output
103	i19	Analog	Inout		LSU Output
104	i27	Analog	Inout		LSU Output
105	i35	Analog	Inout		LSU Output
106	i43	Analog	Inout		LSU Output
107	i51	Analog	Inout		LSU Output
108	clk	Digital	In		Clock input
109	reset	Digital	In		Global reset
110	data_in	Digital	In		Serial input data line
111	prog_en_in	Digital	In		Control line to start to programming the chip
112	stim_en	Digital	In		Control line to enable the stimulation
113	refresh	Digital	In		Control line for activating the refresh of the data
114	data_out	Digital	Out		Daisy chain data out
115	prog_en_out	Digital	Out		Daisy chain progamm enable
116	dvss	Power	Inout		Digital ground
117	dvdd	Power	Inout		1.2 V Digital Power Supply
118	dvssio	Power	Inout		Ground for the IO-Pads
119	dvddio	Power	Inout		3.3 V Supply for the Digital IO-Pads
120		NC			

List of Figures

List of Tables

List of Symbols and Abbreviations

A_{el} .... Area of the electrode-electrolyte interface (electrodes)
B .... Total number of bits for the hybrid DAC
$B1$ .... Number of bits for the unary array
$B2$ .... Number of bits for the binary array
c_m .... Membrane capacitance per unit length [F/m]
C_{blw} .... Inter-lower layers capacitance of skin
C_{bs} .... Inter-stratum corneum capacitance
C_d, C_{dl} .... Double layer capacitance
C_G .... Gouy-Chapman capacitance
C_H .... Helmholtz capacitance
C_{IHP} .... Inner Helmholtz capacitance
C_{lw} .... Lower layers capacitance of skin
C_m , C_{Mem} .... Membrane capacitance
C_{nl} .... Non-linear capacitance of skin
C^o .... Surface concentration of oxidant species in the bulk
C_{OHP} .... Outer Helmholtz capacitance
C^r .... Surface concentration of reductant species in the bulk
C_S .... Stern capacitance
C_s .... Stratum corneum capacitance
C_{Tnl} .... Total non-linear capacitance
$\mathbf{d}$ .... Distance between the electrode and electrolyte
d_{IHL} .... Distance of the inner Helmholtz region
d_{OHL} .... Distance of the outer Helmholtz region
D .... Average value of the diffusion coefficients of the diffusing species
D_o .... Diffusion coefficients of the oxidant
D_r .... Diffusion coefficients of the reductant
E_{he} .... Half-cell potential
$E_{th}(t)$ .... Energy threshold in function of the stimulation time (t)
f .... The activating function for nerve stimulation
f .... Frequency
f_c .... Cut frequency
F .... Faraday constant
g_m .... Membrane unit conductance
G_m .... Membrane conductance
i_o .... Exchange current density
i_m .... Stimulation current
$\mathrm{i}_{Na\,max}$ .... Maximum sodium current per unit length [A/m]
$\mathrm{I_C(t)}$ .... Current on a capacitor in function of time (t)
$I_{i,n}$ .... Intracellular current at node n
$\mathrm{I_R(t)}$ .... Current on a resistor in function of time (t)
$\mathrm{I_S}$ .... Current externally applied
$\mathrm{I_S(t)}$ .... Current externally applied in function of time (t)
$\mathrm{I_{th}(t)}$ .... Current threshold in function of the stimulation time (t)
k .... Boltzmann constant
l .... Length
n .... Number of electrons transferred
n_0 .... Bulk concentration of ions in a solution
N_{Cnl} .... Number of capacitances placed in the model
N_{Rnl} .... Number of resistances placed in the model
q^M .... Excess charge on the electrode surface
q^S .... Excess of cations or anions in the electrolyte
$\mathrm{Q(t)}$ .... Charge in function of time (t)

Q_{inj} .... Charge injection
r .... Radius
r_i .... Axial resistance per unit length [Ω/m]
R .... Gas constant
R_{axp}.................. Axon electrical resistance
R_{blw}.................. Inter-lower layers resistance of skin
R_{bm} .... Inter-biceps Brachii muscle electrical resistance
R_{bs} .... Inter-stratum corneum electrical resistance
R_{ct}.................... Charge transfer resistance
R_d, R_{dl} .... Double layer resistance
R_i .... Intracellular electrical resistance
R_{lw} .... Lower layers resistance of skin
R_m..................... Biceps Brachii muscle electrical resistance
R_m , R_{Mem} .... Membrane electrical resistance
R_{nl}.................... Non-linear resistance of skin
R_s .... Spreading electrical resistance
R_s .... Stratum corneum electrical resistance
S........................ Total number of current cells for hybrid DAC
T........................ Absolute temperature
v........................ Velocity of the action potential
V_0...................... Potential applied to the electrode
$V_c(t)$.................. Voltage on a capacitor in function of time (t)
V_{DS} .... Drain-Source voltage
$V_{DS,sat}$............... Drain-Source saturation voltage
$V_{e,n}$.................. Extracellular potential at node n
V_{GS} .... Gate-Source voltage
$V_{GS,sat}$............... Gate-Source saturation voltage
$V_{i,n}$ .... Intracellular potential at node n
$V_m(t)$.................. Membrane voltage in function of time (t)
$VMODEL_{im}$ .... Voltage simulated in function of the current (i_m) applied
V_r, V_{rp} .... The resting membrane potential
$VTENS_{im}$........... Voltage measured in TENS experiment in function of the current (i_m) applied
V_{th}.................... Membrane threshold voltage
V_{THN}................. Transistor saturation voltage
W .... Warburg element
Z........................ Impedande
Z_W .... Impedance of the Warburg element
β .... MOS transconductance coefficient
δ........................ Nernst diffusion layer thickness
Δ........................ Time step
ϵ........................ Absolute permittivity
ϵ_o .... Vacuum permittivity
ϵ_{IHL} .... Permittivity of the inner Helmholtz region
ϵ_{OHL}.................. Permittivity of the outer Helmholtz region
ϵ_r....................... Relative permittivity
ρ........................ Resistivity of the material
σ........................ Conductivity
σ_{INL} .... Standard deviation of hybrid DAC
σ_E .... Element matching associated error of hybrid DAC
σ^M .... Excess charge on the electrode surface
σ_W .... Warburg coefficient
τ........................ Time constant
Ag .... Silver
AgCl................. Silver Chloride

ASIC .... Application Specific Integrated Circuit
CNS .... Central Nervous System
CPE................... Constant Phase Element
CV..................... Cyclic Voltammetry
CMOS .... Complementary Metal Oxide Semiconductor
DAC .... Digital to Analog Converter
DBS .... Deep Brain Stimulation
DNL................... Differential Non Linearity
ECP................... Electrically Conducting Polymer
EDOT .... EthyleneDiOxyThiophene
EIS .... Electrochemical Impedance Spectroscopy
EMS .... Electrical Muscle Stimulation
ESA................... Electrochemical/electroactive Surface Area
FFT .... Fast Fourier Transform
FNS................... Functional Neuromuscular Stimulation
G .... Gravity
GSA .... Geometric Surface Area
GUI.................... Graphical User Interface
HVLDMOS .... High-Voltage-Laterally-Diffused-Metal-Oxide-Semiconductor
INL .... Incremental Non Linearity
IHP .... Inner Helmholtz Plane
K+ .... Potassium
KCl.................... Potassium Chloride
KSPS .... Kilo Samples Per Second
LGN .... Lateral Geniculate Nucleous
LSB................... Least Significant Bits
LSU................... Local Stimulation Unit
MOS.................. Metal Oxide Semiconductor
MSB.................. Most Significant Bit
Na+ .... Sodium
NaCl.................. Sodium Chloride
NaPSS .... Sodium PolyStyreneSulfonate
NMOS .... N-type Metal Oxide Semiconductor
NMES................ Neuromuscular Electrical Stimulation
OHP .... Outer Helmholtz Plane
OTA .... Operational Transconductance Amplifier
PNS................... Peripheral Nervous System
PMOS .... P-type Metal Oxide Semiconductor
Pt .... Platinum
PtIr.................... Platinum Iridum
PBT................... PolyBiThiophene
PEDOT............. Poly-EthyleneDiOxyThiophene
PPy .... PolyPyrrole
PSS................... PolyStyreneSulfonate
RMS .... Root Mean Square
SFDR .... Spurious Free Dynamic Range
SPS................... Samples Per Second
TENS .... Transcutaneous Electrical Nerve Stimulation
VNS................... Vagus Nerve Stimulation
VHDL .... Very high speed integrated circuit Hardware Description Language

References

[1] P. H. Peckham, "Principles of Electrical Stimulation," *Top. Spinal Inj. Rehabil*, vol. 5, no. 1, pp. 1-5, 1999.

[2] S. Thanos, P. Heiduschka and T. Stupp, "Implantable visual prostheses," *Acta Neurochir. Suppl.*, vol. 97, no. 2, pp. 465-472, 2007.

[3] J. N. Burghartz and e. al., "CMOS imager technologies for biomedical applications," *ISSCC,* 2008.

[4] G. Heinz-Gerd and e. al., "High Dynamic Range CMOS Imager Technologies for Biomedical Applications," *IEEE JSSC,* 2008.

[5] A. Y. Chow and e. al., "The artificial silicon retina microchip for the treatment of vision loss from retinitis pigmentosa," *Arch Ophthalmol,* vol. 122, pp. 460-469, 2004.

[6] S. Klauke and e. al., "Stimulation with a wireless intraocular epiretinal implant elicits visual percepts in blind humans," *IOVS,* 2010.

[7] C. Veraart and e. al., "Pattern Recognition with the Optic Nerve Visual Proshtesis," *Artif. Organs.,* vol. 27, no. 11, 2003.

[8] C. Veraart and e. al., "Visual Sensations produced by optic nerve stimulation using an implanted self-sizing sprial cuff electrode," *Brain Research,* vol. 813, pp. 181-186, 1998.

[9] J. S. Pezaris and R. C. Reid, "Demonstration of artificial visual percepts generated through thalamic microstimulation," *PNAS,* vol. 104, no. 18, pp. 7670-7675, 2007.

[10] W. H. Dobelle, "Artificial Vision for the Blind by Connecting a Television Camera to the Visual Cortex," *ASAIO Journal,* vol. 46, pp. 3-9, 2000.

[11] J. T. Rubinstein, "How cochlear implants encode speech," *Current Opinion in Otolaryngology & Head and Neck Surgery,* vol. 12, pp. 444-448, 2004.

[12] A. H. Milby, C. Halpern and G. Baltuch, "Vagus nerve stimulation in the treatment of refactory epilepsy," *Neurotherapeut,* vol. 6, no. 2, pp. 228-237, 2009.

[13] A. J. Rush and e. al., "Vagus Nerve Stimulation (VNS) for Treatment-Resistant Depressions: A Multicenter Study," *Biol Psychiatry,* vol. 47, pp. 276-286, 2000.

[14] M. B. Shapiro and e. al., "Effects of STN DBS on rigidity in Parkinson´s disease," *IEEE Transactions On Neural Systems and Rehabilitation Engineering,* vol. 15, 2007.

[15] T. L. Skarpaas and M. J. Morrell, "Intracranial stimulation therapy for epilepsy," *Neurotherapeut,* vol. 6, no. 2, pp. 238-243, 2009.

[16] T. E. Schlaepfer and e. al., "Deep brain stimulation to reward circuitry Alleviates Anhedonia in Refactory Major Depression," *Neuropsychopharmacology,* 2007.

[17] I. U. Isaias, R. L. Alterman and M. Tagliati, "Deep brain stimulation for Primary Generalized Dystonia," *Archneurol,* vol. 66, 2009.

[18] G. Barolat and A. D. Sharan, "Spinal cord stimulation for chronic pain managment," *Seminars in neurosurgery,* vol. 15, no. 2/3, 2004.

[19] P. H. Peckham, "Functional Neuromuscular Stimulation," *Phys. Technol.,* vol. 12, 1981.

[20] E. A. Pohlmeyer and e. al., "Toward the restoration of hand use to a paralyzed monkey: brain-controlled functional electrical stimulation of forearm muscles," *PLOS ONE,* vol. 46, no. 6, 2009.

[21] E. A. Pohlmeyer and e. al., "Real-time control of the hand by intracortically controlled functional neuromuscular stimulation," *IEEE ICORR,* 2007.

[22] L. A. Johnson and A. J. Fuglevand, "Mimicking muscle activity with electrical stimulation," *J. Neural Eng.,* vol. 8, 2011.

[23] H. Ring and N. Rosenthal, "Controlled Study of Neuroprosthetic Functional Electrical Stimulation in Sub-Acute Post-Stroke Rehabilitation," *J. Rehabil. Med.,* vol. 37, pp. 32-36, 2005.

[24] J. Chae, L. Sheffler and J. Knutson, "Neuromuscular Electrical Stimulation for Motor restoration in Hemiplegia," *Topics in Stroke Rehabilitation,* vol. 15, no. 5, pp. 412-426, 2008.

[25] N. A. Maffiuletti, M. A. Minetto, D. Farina and R. Bottinelli, "Electrical stimulation for neuromuscular testing and training: state-of-the art and unresolved issues," *European Journal of Applied Physiology,* pp. 2391-2397, 2011.

[26] S. A. P. Haddad, R. P. M. Houben and W. A. Serdijin, "The evolution of Pacemakers," *IEEE Eng. Med. Biol. Mag.,* 2006.

[27] A. Ba and M. Sawan, "Integrated programmable neurostimulator to recuperate the bladder functions," *IEEE CCECE - CCGEI,* vol. 1, pp. 147-150, 2003.

[28] V. Lin and I. N. Hsiao, "Functional Neuromuscular Stimulation of the Respiratory Muscles for Patients With Spinal Cord Injury," *IEEE,* vol. 96, no. 7, 2008.

[29] A. F. DiMarco and e. al., "Phrenic nerve pacing via intramuscular diaphragm electrodes in tetraplegic subjects," *CHEST,* vol. 127, pp. 671-678, 2005.

[30] C. Van Hoof and e. al., "Design and Integration Technology for Miniature Medical Microsystems," *IEEE IEDM,* 2008.

[31] Y. K. Song and e. al., "Microelectronic neurosensor arrays: Towards implantable brain communication interfaces," *IEEE IEDM,* 2008.

[32] P. Fromherz, "Electronic and ionic devices: Semiconductor chips with brain tissue," *IEEE IEDM,* 2008.

[33] K. E. Barrett, S. M. Barman, S. Boitano and H. L. Brooks, Ganong´s Review of Medical Physiology, McGraw-Hill, 2010.

[34] J. Malmivuo and R. Plonsey, Bioelectromagnetism - Principles an Applications of Bioelectric and Biomagnetic Fields, New York: Oxford University Press, 1995.

[35] W. L. Rutten, "Selective electrical interfaces with the nervous system.," *Annual Review of Biomedical Engineering,,* vol. 4, no. 1, pp. 407-452, 2002.

[36] A. Hodgkin and A. Huxley, "A Quantitative Description of Membrane current and its Application to Conduction and Excitation in Nerve," *J Physiol.,* vol. 117, no. 4, pp. 500-544, 1952.

[37] M. Satyam and K. Ramkumar, Foundations of Electronic Devices, New Age International, 1990.

[38] P. H. Rieger, Electrochemistry, 2nd Edition ed., Springer Netherlands, 1994. © 1994 Springer. With kind permission from Springer Science+Business Media.

[39] S. Srinivasan, Fuel Cells From Fundamentals to Applications, Springer, 2006.

[40] A. J. Bard, G. Inzelt and F. Scholz, Electrochemical Dictionary, 2nd Edition ed., Springer Science & Business Media, 2012.

[41] D. C. Grahame, "The Electrical Double Layer and the Theory of Electrocapillarity," *Chemical Reviews,* vol. 41, pp. 441-501, 1947.

[42] W. Franks, I. Schenker, P. Schmutz and A. Hierlemann, "Impedance Characterization and Modeling of Electrodes for Biomedical Applications," *IEEE TBME,* vol. 52, no. 7, 2005.

[43] V. F. Lvovich, Impedance Spectroscopy: Applications to Electrochemical and Dielectric Phenomena, Wiley, 2012.

[44] E. Barsoukov and J. R. Macdonald, Impedance Spectroscopy: Theory, Experiment, and Applications, 2nd ed., Wiley, 2005.

[45] S. F. Cogan, "Neural Stimulation and Recording Electrodes," *Annu. Rev. biomed. Eng.,* vol. 10, pp. 275-309, 2008. “Reproduced with permission of Annual Review of Biomedical Engineering, Volume 10 ©2008 by Annual Reviews, http://www.annualreviews.org".

[46] D. R. Merrill, M. Bikson and J. G. Jefferys, "Electrical Stimulation of Excitable Tissue: Design of Efficacious and Safe Protocols," *Journal of Neuroscience Methods,* vol. 141, pp. 171-198, 2005. Reprinted with permission from Elsevier ©2005.

[47] S. F. Cogan, P. R. Troyk, J. Ehrlich and T. Plante, "In vitro comparison of the charge-injection limits of activated iridium oxide (AIROF) and platinum-iridium microelectrodes," *IEEE Trans. Biomed. Eng.,* vol. 52, pp. 1612-1614, 2005.

[48] T. Rose and L. Robblee, "Electrical stimulation with Pt electrodes. VIII. Electrochemically safe charge injection limits with 0.2 ms pulses," *IEEE Trans. Biomed. Eng.,* no. 37, pp. 1118-1120, 1990.

[49] X. Cui, J. Wiler, M. Dzaman, R. A. Altschuler and D. C. Martin, "In Vivo Studies of Polypyrrole/Peptide coated neural probes," *Elsevier Biomaterials,* vol. 24, pp. 777-787, 2003.

[50] M. Bongo, O. Winther-Jensen, S. Himmelberger, X. Strakosas, M. Ramuz, A. Hama, E. Stavrinidou, G. G. Malliaras, A. Salleo, B. Winther-Jensen and R. M. Owens, "PEDOT:Gelatin Composites mediate brain endothelial cell adhesion," *J. Mater. Chem. B.,* vol. 1, pp. 3860-3867, 2013.

[51] K.-P. Hoffmann, R. Ruff and W. Poppendieck, "Long-Term Characterization of Electrode Materials for Surface Electrodes in Biopotential Recording," *IEEE EMBS,* vol. 28, 2006.

[52] W. Poppendieck and K. Hoffmann, "Coating of neural microelectrodes with intrinsically conducting polymers as a means to improve their electrochemical properties," *ECIFMBE Proceedings,* vol. 22, pp. 2409-2412, 2008.

[53] A. Kros, N. A. Sommerdijk and R. J. Nolte, "Poly(pyrrole) versus poly(3,4-ethylenedioxythiophene): implications for biosensor applications," *Sensors and Actuators B. Chem,* vol. 106, pp. 289-295, 2004.

[54] X. Cui and D. C. Martin, "Electrochemical deposition and characterization of poly(3,4-ethylenedioxythiophene) on neural microelectrode arrays," *Sensor Actuat B-Chem,* vol. 89, pp. 92-102, 2003.

[55] S. M. Richardson-Burns, J. L. Hendricks and D. C. Martin, "Electrochemical polymerization of conducting polymers in living neural tissue," *J. Neural Eng.,* vol. 4, 2007.

[56] S. J. Wikls, S. M. Richardson-Burns, J. L. Hendricks, D. C. Martin and K. J. Otto, "Poly(3,4-ethylenedioxyophene) as a micro-neural interface material for electrostimulation," *Frontiers in Neuroengineering,* vol. 2, 2009.

[57] C. Hassler and T. Stieglitz, "Polymer-Based Approaches to Improve the Long Term Performance of Intracortical Neural Interfaces," *IFMBE Proceedings,* pp. 119-122, 2009.

[58] X. T. Cui and D. D. Zhou, "Poly (3,4-Ethylenedioxythiophene) for Chronic Neural Stimulation," *IEEE TNSRE,* vol. 15, no. 4, 2007.

[59] S. Venkatraman, J. Hendricks, S. Richardson-Burns, E. Jan, D. Martin and J. M. Carmena, "PEDOT coated Microelectrode Arrays for Chronic Neural Recording and Stimulation," in *IEEE EMBS Conference on Neural Engineering,* 2009.

[60] A. M. Nardes, On the conductivity of PEDOT:PSS thin films, Technische Universiteit Eindhoven, 2007.

[61] S. Timpanaro, M. Kemerink, F. Towslager, M. De Kok and S. Schrader, "Morphology and conductivity of PEDOT/PSS films studied by scanning-tunneling microscopy," *Chemical Physics Letters,* vol. 394, pp. 339-343, 2004.

[62] P. G. Taylor, J.-K. Lee, A. A. Zakhidov, M. Chatzichristidi, H. H. Fong, J. A. DeFranco, G. G. Malliaras and C. K. Ober, "Orthogonal Patterning of PEDOT:PSS for Organic Electronics using Hydrofluoroether Solvents," *Adv. Mater,* vol. 21, pp. 2314-2317, 2009.

[63] T. Osaka, S. Komaba and T. Momma, "Conductive Polymers: Electroplating of Organic Films," in *Modern Electroplating,* John Wiley & Sons, 2010, pp. 421-432.

[64] F. Beck, "Electrodeposition of polymer coatings," *Electrochimica Acta,* vol. 33, no. 7, pp. 839-850, 1988.

[65] D. D. Zhou, X. T. Cui, A. Hines and R. J. Greenberg, "Conducting Polymers in Neural Stimulation Applications," in *Implantable Neural Prostheses 2, Techniques and Engineering Approaches,* Springer, 2009. ©2009 Springer. With kind permission from Springer Science+Business Media.

[66] G. Schopf and G. Kossmehl, Polythiophenes - Electrically Conductive Polymers, Springer, 1997. ©1997 Springer. With kind permission from Springer Science+Business Media.

[67] B. Onaral and H. Schwan, "Linear and nonlinear properties of platinum electrode polarisation II: time domain analysis," *Med. and Biol. Eng. and Comput.,* pp. 210-216, 1983.

[68] R. Starbird, W. Bauhofer, M. Meza-Cuevas and W. H. Krautschneider, "Effect of experimental factors on the properties of PEDOT-NaPSS galvanostatically deposited from an aqueous micellar media for invasive electrodes," in *BMEICON, IEEE,* 2012. ©2012 IEEE. Reprinted, with permission.

[69] R. S. Perez, Study of Organic Materials to Improve the Electrical Properties of the Neural Stimulation Electrodes, Dr. Hut, 2013.

[70] P. Manisankar, C. Vedhi, G. Selvanathan and H. Prabu, "Influence of surfactants on the electrochromic behavior of poly (3,4-ethylenedioxythiophene)," *Journal of Applied Polymer Science,* vol. 104, pp. 3285-3291, 2007.

[71] N. Sakmeche, J. Aaron, S. Aeiyach and P. Lacaze, "Usefulness of aqueous anionic micellar media for electrodeposition of poly-(3,4-ethylenedioxythiophene) films on iron, mild steel and aluminium," *Electrochimica Acta,* vol. 45, pp. 1921-1931, 2000.

[72] S. Jezernik and T. Sinkjaer, "Finite element modeling validation of energy-optimal electrical stimulation waveform," *IFESS,* 2005.

[73] C. Robillard, J. Coulombe, P. Nadeau and M. Sawan, "Neural stimulation safety and energy efficiency: Waveform analysis and validation," *IFESS,* 2006.

[74] M. Sahin and Y. Tie, "Non-rectangular waveforms for neural stimulation with practical electrodes," *J. Neural Eng,* vol. 4, pp. 227-233, 2007.

[75] S. D. Bennie, J. S. Petrofsky, J. Nisperos, M. Tsurudome and M. Laymon, "Toward the optimal waveform for electrical stimulation of human muscle," *Eur. J. Appl Physiol,* vol. 88, pp. 13-19, 2002.

[76] M. E. Halpern, "Current waveforms for neural stimulation-charge delivery with reduced maximum electrode voltage," *IEEE TBME,* vol. 57, no. 9, 2009.

[77] C. C. McIntyre and W. M. Grill, "Selective Microstimulation of Central Nervous System Neurons," *Annals of Biomedical Engineering,* vol. 28, pp. 219-233, 2000.

[78] C. C. McIntyre and W. M. Grill, "Extracellular Stimulation of Central Neurons: Influence of Stimulus Waveforms and Frequency on Neuronal Output," *J. Neurophyisiol,* vol. 88, pp. 1592-1604, 2002.

[79] T. Ma, Y.-Y. Gu and Y.-T. Zhang, "Circuit Models For Neural Information Processing," in *Neural Engineering*, Springer US, 2005, pp. 333-365.

[80] D. Luján Villareal, Equivalent Circuit Model to Simulate the Neuromuscular Electrical Stimulation, Technische Universität Hamburg-Harburg, 2010.

[81] D. Lujan Villareal, "Equivalent circuit model to simulate the neuromuscular electrical stimulation," in *ICT.Open*, 2012.

[82] D. Lujan Villareal, "Equivalent Circuit Model to Simulate Neurostimulation by using Different Waveforms," in *ICT.Open*, 2013.

[83] Z. Qi and G. Qingzhi, "A New Kinetic Equation for Intercalation Electrodes," *Jornal of the Electrochemical Society,* 2006.

[84] A. A. o. O. Surgeons, Emergency Care and Transportation of the Sick and Injured, Jones & Bartlett Learning, 2006.

[85] S. Dorgan and R. Reilly, "A Model for Human Skin Impedance During Surface Functional Neuromuscular Stimulation," *IEEE Transactions on Rehabilitation Engineering,* 1999.

[86] A. Van Boxtel, "Skin Resistance During Square-Wave Electrical Pulses of 1 to 10 mA," *med. & Biol. Eng. & Comput.,* 1977.

[87] A. Y. Chan, Biomedical Device Technology: Principles and Design., Charles C Thomas Publisher, 2008.

[88] G. Sverre and Ø. G. Martinsen, Bioimpedance and Bioelectricity Basics, Academic Press, 2008.

[89] D. Miklavcic, N. Pavselj and H. Francis, Electric Properties of Tissues, Wiley Encyclopedia of Biomedical Engineering, 2006.

[90] T. Yamamoto and Y. Yamamoto, Electrical Properties of the Epidermal Stratum Corneum, Medical and Biological Engineering, 1976.

[91] J. Petrofsky, "The Effect of the Subcutaneous Fat on the Transfer of Current Through Skin and into Muscle," *Medical Engineering & Physics,,* 2008.

[92] T. Yamamoto and Y. Yamamoto, "Non-Linear Electrical Properties of Skin in the Low Frequency Range.," *Med. & Bio. Eng. & Comput.,* 1981.

[93] J. G. Marks and J. J. Miller, Lookingbill and Marks's Principles of Dermatology, Elsevier, 2006.

[94] A. Douglas and C. Christensen, Electrical Properties of the Human Body. Basic Introduction to Bioelectromechanics, CRC Press, 2009.

[95] A. Kuhn, Modeling Transcutaneous Electrical Stimulation, Diss. ETH No. 17948, 2008.

[96] M. Gomis Bataller, Efectos del Entrenamiento con Electroestimulacion Muscular en Pacientes Afectados de Hemofilia A, Servei de Publicacions Universitat de Valencia, 2008.

[97] D. R. McNeal, "Analysis of a Model for Excitation of Myelinated Nerve," *IEEE Transactions on Biomedical Engineering,* vol. 23, no. 4, 1976.

[98] J. H. Meier, W. L. Rutten, A. E. Zoutman, H. B. K. Boom and P. Bergveld, "Simulation of Multipolar Fiber Selective Neural Stimulation Using Intrafascicular Electrodes," *IEEE Trans. biomed. Eng.,* vol. 39, no. 2, pp. 122-134, 1992.

[99] C. Koch, "Cable Theory in Neurons with Active, Linearized Membranes," *Bol. Cybern.,* vol. 50, pp. 15-33, 1984.

[100] U. Moran, R. Phillips and R. Milo, "SnapShot: Key Numbers in Biology," *Cell Elsevier,* vol. 141, no. 7, 2010.

[101] R. Hobbie and B. Roth, Intermediate Physics for Medicine and Biology, Springer Science & Business Media, 2007.

[102] H. Ye, M. Cotic, E. E. Kang, M. G. Fehlings and P. L. Carlen, "Transmembrane Potential Induced on the Internal Oranelle by a Time-Varying Magnetic Field: A Model Study," *Journal of NeuroEngineering and Rehabilitation,* vol. 7, no. 12, 2010.

[103] M. A. Meza-Cuevas, A. Gummenscheimer, D. Schroeder and W. H. Krautschneider, "Portable Neuromuscular Electrical Stimulator," in *ICT.Open*, Rotterdam, Netherlands, 2012.

[104] M. A. Meza-Cuevas, D. Schroeder and W. H. Krautschneider, "Neuromuscular Electrical Stimulation Using Different Waveforms: Properties Comparison by Applying Single Pulses," in *IEEE BMEI*, 2012. ©2012 IEEE. Reprinted, with permission.

[105] M. Meza, L. Abu-Saleh, D. Schroeder and W. Krautschneider, "ASIC for Neurostimulation with Different Waveforms," in *BMT*, 2011.

[106] M. A. Meza Cuevas, L. Abu Saleh, D. Schroeder and W. Krautschneider, "Toward the Optimal Architecture of an ASIC for Neurostimulation," in *BIODEVICES*, Vilamoura, Portugal, 2012.

[107] M. A. Meza-Cuevas, K. Ramesh, D. Schroeder and W. H. Krautschneider, "Hybrid Architecture of a DAC for Neurostimulation," in *BMEICON, IEEE*, Ubon Ratchathani, Thailand, 2012. ©2012 IEEE. Reprinted, with permission.

[108] M. A. Meza-Cuevas, D. Schroeder and W. H. Krautschneider, "A Scalable 64 Channel Neurostimulator based on a Hybrid Architecture of Current Steering DAC," in *IEEE MECBME*, Doha, Qatar, 2014. 2014 IEEE. Reprinted, with permission.

[109] R. J. Baker, CMOS Circuit design, Layout and Simulation, IEEE Press Series on Microelectronic Systems, 2005.

[110] D. A. Johns, Analog Integrated Circuit Design, John Wiley & Sons, 1997.

[111] B. A. Minch, "A low-voltage MOS cascode current mirror for all curren levels," *MWSCAS,* 2002.

[112] S. Ethier and M. Sawan, "Exponential Current Pulse Generation for Efficient Very High-Impedance Multisite Stimulation," *IEEE TBCAS,* vol. 5, pp. 30-38, 2010.

[113] R. Ramasamy, B. Venkataramani, C. Rajkumar, B. Prashanth and K. K. Bharath, "The design of an area efficient segmented DAC," *IEEE ICSIP,* 2010.

[114] J. D. Weiland and e. al., "Systems design of a high resolution retinal prosthesis," *IEEE IEDM,* 2008.

[115] K. Cha, K. Horch and R. Normann, "Stimulation of a Phosphene-Based Visual Field: Visual Acuity in a Pixelized Vision System," *Annals of Biomedical engineering,* vol. 20, pp. 439-449, 1992.

[116] R. Shepherd, N. Linahan, J. Xu, G. M. Clark and S. Araki, "Chronic Electrical Stimulation of the Auditory Nerve using Non-charge-balanced Stimuli," *Acta Otolaryngol,* vol. 119, pp. 674-684, 1999.

[117] E. Noorsal, K. Sooksood, H. Xu, R. Hornig, J. Becker and M. Ortmanns, "A Neural Stimulator Frontend With High-Voltage Compliance and Programmable Pulse Shape for Epiretinal Implants," *IEEE Journal of Solid-State Circuits,* vol. 47, no. 1, 2012.

[118] N. Tran, E. Skafidas, J. Yang, S. Bai, M. Fu, D. Ng, M. Halpern and I. Mareels, "A Protoype 64-Electrode Stimulator in 65 nm CMOS Process towards a High Density Epi-retinal Prosthesis," in *IEEE EMBS Annual International Conference*, 2011.

Curriculum Vitae

Surname	Meza Cuevas
Name	Mario Alberto
Birthday	September 5th, of 1982
Birthplace	Guadalajara, Jalisco, Mexico

09.1994-08.1997	Technical Middle School No. 1 of Guadalajara Technical Middle School in Accounting
09.1998-08.2001	Technical High School No. 11 of University of Guadalajara Technical High School in Cytology and Histology
09.2001-08.2007	CUCEI, University of Guadalajara Engineering degree in electronic and communications Final work: "Traffic Light with Priority Pass for Ambulances"
09.2007-11.2009	CINVESTAV Campus Guadalajara Master of Science oriented to electronic design Thesis project: "Design of an OCDMA Communication System Between Integrated Circuits"
01.2010-11.2015	Institute of Nanoelectronics (now is called Institute of Nano- and Medicine-Electronics) at Hamburg University of Technology PhD in nanoelectronics for medical applications Dissertation: "Stimulations of Neurons by Electrical Means"

09.1995-04.1996	Pont Computers / Guadalajara, Mexico Technician
05.1996-05.2006	Maxcom / Guadalajara, Mexico Owner Shop of sell, reparation and maintenance of computer equipment
07.2005-08.2005	Acterna Deutschland GmbH / Eningen, Germany Internship Departments of assembly, testing, design, development and support
01.2007-08.2007	Private Sector / Guadalajara, Mexico Design, develop and implementation of project Project to automate a process of label collocation in a plastic moldering machine, using a robotic arm and electrostatic charge
01.2009-11.2009	Cuauhtémoc University Campus Guadalajara / Guadalajara, Mexico Bachelor Teacher Teaching the courses: "Electronics I and II" and "Electric Circuits I and II"
11.2011-11.2012	Institute of Nanoelectronics at Hamburg University of Technology Temporary employee (Aushilfskraft) TuTech "Stimulation von Muskeln mit elektrische Spannungspulsen"
12.2012-11.2013	Institute of Nanoelectronics at Hamburg University of Technology Research Associate (Wissenschaftlicher Mitarbeiter) BMWi "Exist-Forschungstransfer Projekt" Project: BioParSys
12.2013-02.2014	Institute of Nanoelectronics at Hamburg University of Technology Research Associate (Wissenschaftlicher Mitarbeiter) BMBF "MyoPlant"
Since 11.2013	Hamburg Applications MES UG / Hamburg, Germany Co-owner Company developing Medical Electronic Systems
Since 05.2014	NXP Semiconductors / Hamburg, Germany Contractor Design of a security test chip for smart card interfaces

Bisher erschienene Bände der Reihe

Wissenschaftliche Beiträge zur Medizinelektronik

ISSN 2190-3905

1	Wjatscheslaw Galjan	Leistungseffiziente analogdigitale integrierte Schaltungen zur Aufnahme und Verarbeitung von biomedizinischen Signalen ISBN 978-3-8325-2475-3 37.50 EUR
2	Jakob M. Tomasik	Low-Noise and Low-Power CMOS Amplifiers, Analog Front-Ends, and Systems for Biomedical Signal Acquisition ISBN 978-3-8325-2481-4 37.00 EUR
3	Fabian Wagner	Automatisierte Partitionierung von Mixed-Signal Schaltungen für die Realisierung von Systems-in-Package ISBN 978-3-8325-3424-0 38.00 EUR
4	Mario A. Meza-Cuevas	Stimulation of Neurons by Electrical Means ISBN 978-3-8325-4152-1 45.00 EUR
5	Lait Abu Saleh	Implant System for the Recording of Internal Muscle Activity to Control a Hand Prosthesis ISBN 978-3-8325-4153-8 44.50 EUR

Alle erschienenen Bücher können unter der angegebenen ISBN-Nummer direkt online (http://www.logos-verlag.de) oder per Fax (030 - 42 85 10 92) beim Logos Verlag Berlin bestellt werden.